AS-G-III-2
U-X
Z-183

Berliner geographische Studien

Herausgeber: Frithjof Voss

Schriftleitung: Michael Wiesemann-Wagenhuber

Band 46

Satellite Data Analysis and Surface Modeling for Land Use and Land Cover Classification in Thailand

von Kankhajane Chuchip

D 83

Berlin 1997

Institut für Geographie der Technischen Universität Berlin

Geographisches Institut der Universität Kiel ausgesonderte Dublette

Die Arbeit wurde am 17. Dezember 1996 vom Fachbereich Umwelt und Gesellschaft der Technischen Universität Berlin unter dem Vorsitz von Prof. Dr. Lutz Lehmann, Berlin, aufgrund der Gutachten von Prof. Dr. F. Voss, Berlin, Prof. Dr. W. Werner, Wiesloch, und Prof. Dr. K-P. Lade, Salisbury, USA, als Dissertation angenommen.

Herausgeber: Prof. Dr. Frithjof Voss

Schriftleiter: Kankhajane Chuchip

Titelseite: Dipl.-Ing. Hans-Joachim Nitschke
Inset: Kankhajane Chuchip

ISSN 0341-8537
ISBN 3 7983 1706 2

Gedruckt auf säurefreiem alterungsbeständigem Papier

Druck/ Printing: Offset-Druckerei Gerhard Weinert GmbH
Saalburgstraße 3, D-12099 Berlin - Tempelhof

Vertrieb/ Publisher: Technische Universität Berlin
Universitätsbibliothek, Abt. Publikationen
Straße des 17. Juni 135, D-10623 Berlin - Charlottenburg
Tel.: (030) 314-22976, -23676
Fax.: (030) 314-24743

Geographisches Institut
der Universität Kiel

Verkauf/ Book-Shop: Gebäude FRA-B
Franklinstraße 15 (Hof), D-10587 Berlin - Tiergarten

Inv.-Nr. A 28175
Geographisches Institut
der Universität Kiel
ausgesonderte Dublette

ACKNOWLEDGEMENT

I wish to acknowledge the support and assistance of many people. Without them my research would not have been accomplished easily.

In particular, I am indepted to Prof. Dr. Frithjof Voss who gave me a chance to do this research at the Institut für Geographie, Technische Universität Berlin. I was able to graduate because of his helpful advice, support, and assistance in many matters. I would like to thank Mr. Suwit Ongsomwang who has kindly helped me in various ways. Because of him I had an opportunity to know Professor Voss.

I wish to express my gratitude to Prof. Dr. Wolfgang Werner who kindly read my manuscripts and sat on as a committee in my defense. Extra thank go to Prof. Dr. K-P. Lade, one of my committee, who suggested, trained and supported me precious softwares for doing this research.

I also express my sincere thanks to my friend Frank Torkler who has been very understanding, helpful, and supportive during my study in Germany. I have been fortunate in having many of good colleagues in the institute. I appreciate their suggestions and their sympathy. They also helped me in particular ways. My sincerely thanks also go to Mr. Dirk Lehmann for being willing to spend hours reading my manuscript.

I also wish to thank Dipl.-Ing. Joachim Nitschke for his help in preparing this book. I appreciate his excellence in Cartography and artworks.

I am truly grateful to the Faculty of Forestry, Kasetsart University, Bangkok, in supporting me the budget for doing field works. I also wish to thank the National Research Council of Thailand and its administrators who supported me satellite data and relevant information.

I have some more people to thank for their support and assistance. Mr. Prasit Pian-anurak, Mr. Chayut Suktip, Mr. Chompoo Mahantakasri, Mr. Samrit Yincharoen, and some of my students who helped me in the field work. I truly express my gratitude and thanks for all.

Finally, I especially wish to express my appreciation to the Deutscher Akademischer Austauschdienst (DAAD) for financing me throughout my studies in Germany.

I owe them more than a simple debt of gratitude. I thank them all once again.

Kankhajane Chuchip
Berlin, January 1997

This work is for my parents
and
my Thailand.

CONTENTS

LIST OF FIGURES

LIST OF TABLES

1 INTRODUCTION

Remotely sensed data have been used for preparing information of land use and land cover in form of thematic maps in Thailand for many years. Normally, land use and land cover maps are intensively utilized by many government agencies whose mission involves natural resource management planning. According to the rapid change of natural resources, the need for regular natural resources monitoring has been especially important in Thailand. In addition, there has also been an increasing interest in the use of such data for various purposes. The use of remotely sensed data in carrying out various kinds of studies can also be found.

Traditionally, land use and land cover mapping in Thailand are based on the interpretation of aerial photographs and field survey. In general, aerial photographs, which have been a primary source of spatial information for many years, are superior to satellite imagery in terms of spatial resolution. The major drawback of aerial photographs is, however, that the cost per unit of area is high and that they have to be interpreted manually. The Royal Thai Survey Department is the only government agency that produces and distributes aerial photographs. For security reasons, the distribution of aerial photographs is restricted to government agencies and educational institutes. As a result, aerial photographs, as an available data source, have not been used so widely according to those restrictions. Alternately, there is a satellite receiving station situated in Bangkok. It has the capability of acquiring and processing Landsat, SPOT, MOS-1, JERS-1, and ERS-1 data from a series of satellites. The Thailand Remote Sensing Center (TRSC) serves as a data center and provides applications supporting both national and international requests with having not so many restrictions as aerial photographs. Since high resolution satellite data, such as Landsat TM, SPOT- Panchromatic and XS, seem to be the best of the available remotely sensed data sources, the use of satellite data is intended to provide a better understanding of major development trends in Thailand. However, many existing land use and land cover classification systems are still based on aerial photographs or low resolution satellite data. These systems have been developed for many years. Some systems were compiled for specific programs or purposes and mostly for use with manual (visual) interpretation.

In this study, the available land use and land cover classification systems were gathered and evaluated. Then, a classification system was developed. This system is suited for use with high resolution satellite data, in particular Landsat TM, and based on computer-aided analysis. Furthermore, the study was aimed to prepare selective choices of Landsat TM image processing for the classification of various land use and land cover types, and the proper techniques of computer-based analysis and mapping for Thailand's features. The study was also intended to serve for further studies as a learning tool for the use of remote sensing technology, land use and land cover classifications, and mapping, in the term of computerized resource assessment techniques. Thus, overall techniques are extensively described in this research. Finally, results and techniques derived in the study were used to further demonstrate their applicability in a selected test site in Thailand.

However, the purpose of the study was not intended to be global, but to concentrate on a few key techniques relating to remote sensing and relevant themes so as to establish credible schemes necessary for further studies, in particular involving future routine monitoring of land use and land cover changes. This study was divided into two parts according to the main purposes of the study. The first part is related to the analysis of satellite data and surface modeling for land use and land cover classification in Thailand. The second part is related to the application of study results in a selective topic in order to evaluate and distribute the methodology and results of the research from the first part. Specifically, the main objects of this study can be listed as follows:

(1) To assess and modify the existing land use and land cover classification systems used in Thailand, with particular emphasis on the use of high resolution satellite imagery, as a guide for improving computer-aided classification procedures.

(2) To demonstrate satellite data analysis techniques and classification procedures, including field survey corroboration. These include:

- how to model terrain surfaces in the forms of Digital Elevation Models (DEMs) and Triangulated Irregular Networks (TINs) and how to integrate such models into satellite image processing in order to derive the best understanding and results of the land use and land cover classification.
- how to produce thematic maps, DEMs & TINs as well as their associated data based on computerized processing instead of manual methods.

(3) To investigate and demonstrate the possibility of applying techniques and results derived in the study with the selected test site.

2 REVIEW OF LITERATURE

2.1 Land use and land cover classification systems

2.1.1 LAND USE AND LAND COVER CLASSIFICATION

Information on land use and land cover is required in many aspects of land use planning and policy development, as a prerequisite for monitoring and modeling land use and environmental change, and as a basis for land use statistics at all levels (FAO/UNEP, 1995). The distinction between land use and land cover is fundamental. Previous studies have shown that several concepts of land use and land cover are closely related and are interchangeable. Land use refers to human activities on and in relation to the land, whereas land cover denotes the vegetation and artificial constructions covering the land surface (ESTES et al., 1982; LOELKES et al., 1983; LO, 1986). Land use and land cover data, usually presented in the form of a map, are essential to planners who have to make decisions concerning land resource management (LO, 1986). ANDERSON et al. (1976) developed the land use and land cover classification scheme, as shown in Tab. 1, for use with remotely sensed data by considering the following criteria: (1) a minimum level of interpretation accuracy of at least 85 percent; (2) equal accuracy for different categories; (3) repeatable results; (4) applicability over extensive areas; (5) categorization permitting land cover to be used as surrogate for activity; (6) possibilities for use with remotely sensed data acquired at different times; (7) integration with ground surveyed data or large-scale remotely sensed data possible through the use of sub-categories; (8) aggregation of categories possible; (9) possibility of comparison with future data; (10) multiple uses of land recognizable.

Tab. 1: USGS land use and land cover classification system developed for use with remotely sensed data (Source: ANDERSON et al., 1976).

Level 1	Level 2
1 Urban or built-up land	11 Residential
	12 Commercial and services
	13 Industrial
	14 Transportation, communications, and utilities
	15 Industrial and commercial complexes
	16 Mixed urban or built-up land
	17 Other urban or built-up land
2 Agricultural land	21 Cropland and pasture
	22 Orchards, groves, vineyards, nurseries, and ornamental horticulture areas
	23 Confined feeding operations
	24 Other agricultural land
3 Range land	31 Herbaceous range land
	32 Shrub and brush range land
	33 Mixed range land
4 Forest land	41 Deciduous forest land
	42 Evergreen forest land
	43 Mixed forest land

Tab. 1: (continued)

Level 1	Level 2
5 Water	51 Streams and canals
	52 Lakes
	53 Reservoirs
	54 Bays and estuaries
6 Wetland	61 Forested wetland
	62 Non-forested wetland
7 Barren land	71 Dry salt flats
	72 Beaches
	73 Sandy areas other than beaches
	74 Bare exposed rock
	75 Strip mines, quarries and gravel pits
	76 Transitional areas
	77 Mixed tundra
8 Tundra	81 Shrub and brush tundra
	82 Herbaceous tundra
	83 Bare ground tundra
	84 Wet tundra
	85 Mixed tundra
9 Perennial snow or ice	91 Perennial snow-fields
	92 Glaciers

2.1.2 CLASSIFICATION SYSTEMS IN THAILAND

In 1975, the Department of Land Development designed the classification system based on both land use and land cover by using aerial photographs at a scale of 1: 15,000 and field surveys (WACHARAKITTI, 1982) as shown in Tab. 2. This classification was intended to be a standard that could be applied to remotely sensed data. However, many government agencies as well as some researchers have independently produced and used various classification systems. ONGSOMWANG (1993) pointed out that existing land use classification systems in Thailand were modified or defined from various sources by different government departments to serve varying purposes based on their primary responsibility. OMAKUPT (1978) also provided land use maps of Northern Thailand at a scale of 1: 500,000 by interpreting black and white prints of Landsat-2 imagery. The land use classification consisted of five major categories and seven sub-categories as shown in Tab. 3. A similar manual application for practical land use and land cover mapping was carried out by WACHARAKITTI (1982). A part of Chiang Mai province, northern Thailand, was mapped at a scale of 1: 250,000 using color infrared prints. This classification is shown in Tab. 4. In 1988, the Forest Management Division, Royal Forest Department, classified land use for a forest inventory as shown in Tab. 5. The Office of Agricultural Economics has developed a land use classification system, based on the utilization of land for agricultural purpose, as shown in Tab. 6 and 7. Until 1993 when this study was started, it was found that land use and land cover maps prepared by these organizations are mostly still based on visual interpretation.

Tab. 2: Land use classification system of Thailand in 1975 (Source: WACHARAKITTI, 1982).

Level 1	Level 2	Level 3	Level 4
U. Urban land	U.1 Residential land		
	U.2 Commercial land		
	U.3 Institutional land (Schools, hospitals, parks, temples, etc.)		
	U.4 Transportation land (Railroads yards, roads, airfields, etc.)		
	U.5 Industrial land	U.5.1 Mines and dumps	
		U.5.2 Factories	
A. Agricultural land	A.1 Horticultural land	A.1.1 Truck crop land	
		A.1.2 Ornamental plant gardens	
		A.1.3 Vineyards	
		A.1.4 Other horticultural land	
	A.2 Perennial crop land	A.2.1 Orchards	A.2.1.1 Citrus orchards
			A.2.1.2 Durian orchards
			A.2.1.3 Rambutan orchards
			A.2.1.4 Longan orchards
			A.2.1.5 Litchi orchards
			A.2.1.6 Mango orchards
			A.2.1.7 Custard orchards
			A.2.1.8 Jujube orchards
			A.2.1.9 Peach orchards
			A.2.1.10 Apple orchards
			A.2.1.11 Other orchards
		A.2.2 Rubber plantations	
		A.2.3 Coconut plantations	
		A.2.4 Banana plantations	
		A.2.5 Kapok plantations	
		A.2.6 Sugar palm plantations	
		A.2.7 Oil palm plantations	
		A.2.8 Coffee plantations	
		A.2.9 Tea plantations	
		A.2.10 Miang plantations	
		A.2.11 Mulberry plantations	
		A.2.12 Bamboo plantations	
		A.2.13 Pepper plantations	
	A. 3 Field crop land	A.3.1 Corn fields	
		A.3.2 Sugarcane fields	
		A.3.3 Cassava fields	
		A.3.4 Cotton fields	
		A.3.5 Tobacco fields	
		A.3.6 Pineapple fields	
		A.3.7 Soybean fields	
		A.3.8 Peanut fields	
		A.3.9 Nung bean fields	
		A.3.10 Red bean fields	

Tab. 2: (continued)

Level 1	Level 2	Level 3	Level 4
		A.3.11 Fiber crop fields (Jute, kenaf)	
		A.3.12 Cucurbit fields	
		A.3.13 Castor bean fields	
		A.3.14 Sorghum fields	
		A.3.15 Sesame fields	
		A.3.16 Irish potato fields	
		A.3.17 Sweet potato fields	
		A.3.18 Upland rice fields	
		A.3.19 Other field crop land	
	A.4 Paddy land	A.4.1 Broadcast rice land	A.4.1.1 One broadcast rice only
			A.4.1.2 Double broadcast rice cropping
			A.4.1.3 Irrigated, double or triple cropping including one broadcast rice crop as the main crop in rotation with other crops
		A.4.2 Transplanted rice land	A.4.2.1 Rainfall, one transplanted rice crop only
			A.4.2.2 Irrigated, one transplanted rice cropping
			A.4.2.3 Irrigated, double or triple cropping including one broadcast rice crop as the main crop in rotation with other crops
	A.5 Pasture and range land	A.5.1 Improved pasture	
		A.5.2 Unimproved pasture and range land	
	A.6 Swidden land	A.6.1 Land currently under swidden cultivation	A.6.1.1 Upland rice in association with other crops
			A.6.1.2 Corn or opium in association with other crops
			A.6.1.3 Other swidden crop land
		A.6.2 Bush fallow land (Old clearing)	
F. Forest land	F.1 Dry dipterocarp forest	F.1.1 Undisturbed dry dipterocarp forest	
		F.1.2 Disturbed dry dipterocarp forest	
	F.2 Dry dipterocarp forest with pine	F.2.1 Undisturbed dry dipterocarp forest with pine	
		F.2.2 Disturbed dry dipterocarp forest with Pine	

Tab. 2: (continued)

Level 1	Level 2	Level 3	Level 4
	F.3 Mixed deciduous forest with teak	F.3.1 Undisturbed mixed deciduous forest with teak	
		F.3.2 Disturbed mixed deciduous forest with teak	
	F.4 Mixed deciduous forest without teak	F.4.1 Undisturbed mixed deciduous forest without teak	
		F.4.2 Disturbed mixed deciduous forest without teak	
	F.5 Dry evergreen forest	F.5.1 Undisturbed dry evergreen forest	
		F.5.2 Disturbed dry evergreen forest	
	F.6 Moist evergreen forest	F.6.1 Undisturbed moist evergreen forest	
		F.6.2 Disturbed moist evergreen forest	
	F.7 Hill evergreen forest	F.7.1 Disturbed hill evergreen forest	
		F.7.2 Disturbed hill evergreen forest	
	F.8 Mangrove forest	F.8.1 Undisturbed mangrove forest	
		F.8.2 Disturbed mangrove forest	
	F.9 Bamboo		
	F.10 Forest plantations	F.10.1 Teak plantations	
		F.10.2 Pine plantations	
		F.10.3 Other plantations	
I. Idle land			
W. Water bodies	W.1 Salt pans		
	W.2 Shrimp ponds		
	W.3 Fish ponds		
	W.4 Other water bodies (Lakes, ponds, reservoirs, rivers, etc.)		

Tab. 3: Land use classification scheme for land use mapping in North Thailand by OMAKUPT (1978).

Major Categories	Sub-categories
I Urban	
II Agricultural land	1. Horticultural land
	2. Perennial crop land
	3. Field crop land
	4. Paddy land
	5. Pasture and range land

Tab. 3: **(continued)**

Major Categories	Sub-categories
III Forest land	1. Dense forest 2. Cut forest
IV Water body	
V Miscellaneous land	

Tab. 4: **Chiang Mai land use classification system for use with Landsat imagery at 1: 250,000-scale color IR prints (Source: WACHARAKITTI, 1982).**

Level 1	Level 2	Level 3
01 Urban and built-up land	01 Residential	
	02 Transport, communication and utilities	
	03 Strip and clustered settlement	
	04 Mixed	
	05 Open and other	
02 Agricultural land	01 Permanent	01 Irrigated paddy field
		02 Dry paddy field
		03 Cropland and pasture
		04 Orchards, groves, bush fruits, vineyards, and horticultural areas
		05 Other
	02 Shifting	01 Active
		02 Old clearing
03 Rangeland		
04 Forest land	01 Deciduous	01 Undisturbed
		02 Disturbed
	02 Evergreen (Coniferous and other)	01 Undisturbed
		02 Disturbed
	03 Mixed	
	04 Plantation	
05 Water	01 Streams and rivers	
	02 Lakes	
	03 Reservoirs	
	04 Other	
06 Non-productive land		
	01 Sand	
	02 Rock (Exposed)	

Tab. 5: **Land use classification system for forest inventory in Thailand (Source: BOONYOBHAS, 1988).**

Level 1	Level 2	Inventory code
1 Forest land	1.1 Plantation	-
	1.1.1 Teak plantation	000
	1.1.2 Non-teak plantation	001
	1.2 Tropical evergreen forest	100
	1.3 Hill evergreen forest	110
	1.4 Mixed deciduous forest	-
	1.4.1 Mixed deciduous forest with teak	200

Tab. 5: **(continued)**

Level 1	Level 2	Inventory code
	1.4.1 Mixed deciduous forest with teak	200
	1.4.2 Mixed deciduous forest without teak	210
	1.5 Dry dipterocarp forest	-
	1.5.1 Dry dipterocarp forest (High open)	300
	1.5.2 Dry dipterocarp forest (Scrub)	310
	1.6 Pine forest	800
	1.7 Savannah	900
2 Non-forest land	2.1 Old clearing areas	400
	2.2 Shifting cultivation areas	410
	2.3 Agricultural areas	600
	2.4 Other non-forest areas	610

Tab. 6: **Land utilization classification system in Thailand (Source: OFFICE OF AGRICULTURAL ECONOMICS, 1989).**

Level 1	Level 2
1 Forest land	
2 Farm holding land	2.1 Housing areas
	2.2 Paddy land
	2.3 Under field crops
	2.4 Under fruit trees and tree crops
	2.5 Under vegetable and flowers
	2.6 Grass land
	2.7 Idle land
	2.8 Other land
3 Unclassified land	

Tab. 7: **Land use classification system for use with visualization of satellite imagery** (Source: **OFFICE OF AGRICULTURAL ECONOMICS, 1992).**

1 Paddy fields
2 Field crops
3 Horticultural trees
4 Forest
5 Water bodies
6 Other land

It can be noticed that available classification systems are mostly incompatible with each other. These systems are designed for use with image data derived from the old technologies of remote sensing and mostly based on visual interpretation. The system shown in Tab. 2 seems to be the best one which gives more details and covers nearly all possible land use and land cover types. However, several categories included in this system may not be especially separable from another by analyzing satellite data. For example, it could be anticipated that significant confusion would exist between many categories describing various tree species in the classes of agricultural land. Problems may also be encountered in differentiating between the categories of urban land class. On the other hand, the classification of other systems was oriented to what the developer was interested. For example, the system developed by the Royal Forest Department mainly provided only forest type categories, while the system developed by the Office of Agricultural Economics mainly categorized agricultural lands.

Since all of the existing classification systems have not been standardized, the effort of classifying land use and land cover in Thailand seems to be still going on. Due to the fact that mapping has become automated, based on computer-aided procedures, new researchers may develop their own classification systems without adopting any available systems. Since remote sensing makes it increasingly possible to map and monitor land use and land cover over wide areas, a need for establishing a uniform land use and land cover classification system is important. As a result, the optimization of land use and land cover classification systems were considered to be done in this study.

2.2 Satellite remote sensing for land use and land cover classification

Many studies show that the use of remotely sensed high resolution satellite data is particularly appropriate for land use and land cover mapping (e.g., HÄME, 1984; BENSON & DeGLORIA, 1985; BANNINGER, 1987; CALOZ & BLASER, 1987; KADRO, 1987; SCHARDT, 1987; TØMMERVIK, 1987; WULF & GOOSSENS, 1987; STENBACK & CONGALTON, 1990; ZEFF & MERRY, 1993), land use classification (e.g. ERDIN et al., 1986; MAUSER, 1989) and land cover classification (e.g. UENO et al., 1986). From these studies, it can be noticed that the preparation of thematic maps requires a good understanding of land use and land cover classification systems, the proper use of mapping, and image processing procedures. Thus, a number of previous studies relevant to using satellite remote sensing for land use and land cover classification were studied and selected to refer in this study. Since main purposes of this study are to develop a land use and land classification system that should be used with high resolution remotely sensed data and with computerized classification methods in the environmental features of Thailand, practical techniques from previous studies were firstly tested with our data. Only optimal techniques were then adopted to apply to this study.

In principle, computerized image processing uses digital data and falls into two main categories: image enhancement and image analysis. Descriptively, image enhancement algorithms are applied to remotely sensed data to improve the appearance of an image for human visual analysis, or occasionally for subsequent machine analysis (JENSEN, 1986). Principal Component Analysis (PCA), Band Ratio and Vegetation Index are some examples of image enhancement algorithms. They are widespread techniques used for reducing the number of bands to classify without a significant loss of the original information (CHUVIECO, 1987). Ratio transformations of remotely sensed data can be performed to reduce the effect of shadows, seasonal changes in solar angle and intensity, and fluctuations in sensor-surface geometry caused by topographic orientation (JENSEN, 1986). Band ratio was used with a variety of Landsat TM spectral channel combinations in a study by WALSH et al. (1990). In the study it was found that ratio spectral channels for feature enhancement or merging of ratio spectral channels with raw spectral channels within the classification process could improve land cover differentiation in complex environments. PCA is an important data transformation technique used in remote sensing work with multispectral data or other multidimensional data (LILLESAND & KIEFER, 1994). UENO et al. (1986) also used PCA to reduce the dimension of the spectral bands of Thematic Mapper data. These studies show that there is no significant difference between the classification results of raw Landsat TM data and principal component data, whereas the computer time of the classification procedure for principal component data is much less than that for the raw TM data. MAUSER (1989) indicates that if raw TM data are not normally distributed, the principal component transformation will bring it closer to a normal distribution and make it more suitable for the Maximum-Likelihood Classifier. In addition, many studies indicate that the use of ancillary data, consisting of, for example, elevation data, aspect data, soil, and climatic data, can be incorporated in the classification procedure in several ways. OLSSON (1986) stated that if ancillary data can be included in the analysis of image data, the classification results can be improved significantly. SCHARDT (1987) used the additional information from a Digital Elevation Model and data from different seasons to improve the classification results.

Whether all standard analysis methodologies mentioned above should be adopted to widely use in the different kind of study-site environments in Thailand has not been clear. Thus, the analysis techniques of those studies will also be tested in this study.

2.3 Topographic data in form of Digital Elevation Models (DEMs) and Triangulated Irregular Networks (TINs)

As just implied above, the use of topographic data into satellite data analysis has been found in some previous studies. Normally, land use and land cover delineation from remote sensing products by means of visual interpretation require knowledge of sensor systems, object recognition. Interpreters usually use the characteristics of manmade features or objects to classify land use and land cover types from an image. These characteristics are pattern, shape, color tone, size, texture, shadow, and site association. Furthermore, ancillary data such as topographic data are actually also important since some land use and land cover types, such as shifting cultivation areas, forest types, are associated with the topography of areas. However, topographic data are not directly used in the process of image classification by an interpreter. An interpreter usually determines the characteristics of a terrain and its associations based on his background knowledge and his consideration. Furthermore, many interpreters usually concentrate the whole image used during the manual interpretation. This is different from a computer-aided analysis. Image classification is mostly automated. Interpreter is a computer system. Computer users need not to do so many works in the interpretation process. With computer-aided systems, topographic data can be directly adopted into the image classification to improve a result. Such data can also give more meanings about land use and land cover categories derived from the automated classification. Topographic data derived and stored in computer-based systems are widely known in the terms of TIN and DEM. What are TIN, DEM and their roles can be described as follows:

2.3.1 SURFACE MODELS IN FORM OF TINs and DEMs

TIN is an acronym for Triangulated Irregular Network (while TIN™ usually means an optional software module of ARC/INFO software package). TIN is a primary data structure used in ARC/INFO® software for representing continuous surfaces, especially terrain. A TIN connects a set of irregularly-spaced x, y, z locations. TIN properties are especially useful for representing surfaces that are highly variable, and contain discontinuities and break-lines (ESRI, 1992a). Surface elevation, subsurface elevation and terrain modeling can take advantage of the TIN data. On the other hand, a Digital Elevation Model (DEM) contains a systematic, regularly-spaced sample of x, y, z locations. Surface modeling by means of ERDAS™ software involves the processing and graphic simulation of elevation (or z value) data. It has been used for surface representation and producing topographic-relevant data, such as slope or aspect data, essential for using in GIS modeling. Surface data from the ERDAS™ topographic module can produce a DEM, are also useful for representing theoretical surfaces such as non-point pollution densities, chemical concentrations, and ground water values (ERDAS, 1991a). Since DEMs contain a regularly-spaced, systematic sample of a surface, they are more suited than TINs for calculation of slope, aspect, sun intensity, shaded relief and cut-fill areas.

2.3.2 THE ROLE OF TOPOGRAPHIC DATA IN CONJUNCTION WITH REMOTE SENSING AND THE GEOGRAPHIC INFORMATION SYSTEM

In recent years, there has been an increasing interest in the use of topographic data in form of Digital Elevation Models (DEMs) in many kinds of studies involving application of topographic data and their associated products. It was found that topographic data, in particular elevation, slope, and aspect, are among the most important data in many natural resource spatial databases (BOLSTAD & STOWE, 1994). RITTER (1987) also noted in a similar way that slope and aspect information, alone or in combination with other data, can be of considerable valuc in a Geographic Information System. Similarly, ISAACSON & RIPPLE (1990) reported that improvement of the capability to conduct inventories of natural resources can be achieved by accounting for topographic variables such as slope gradient and slope aspect, and elevation. With GIS, topographic variables such as these can be developed from a DEM, available in digital format with the help of a computerized system. Some valuable examples of the application of DEMs include estimating potential availability of solar energy, determining suitability for land development, determining erosion potential, mapping for forest management, visibility analysis, three-dimensional display of landforms for various purposes, and route design (BURROUGH, 1986; RITTER, 1987). A DEM can be used as a background for displaying thematic information or for combining relief data with thematic data such as soils, land use or vegetation. There are a number of studies relating to integrating DEM data with remotely sensed data. SCHARDT (1987) applied DEM data to his study, in which TM data were used for forest classification in the area of Freiburg, Germany. In the study, a DEM was integrated as the additional information to avoid misclassifications caused by the influence of topographic differences between the forest of the Rhine plain and the higher areas of the Kaiserstuhl. LEPRIEUR & DURAND (1988) also used DEMs and Landsat data to study the influence of topography on forest reflectance. In this study, a DEM was compiled with a resolution of 30 meters to study the relationship between reflectance, the variable of interest, and the local sun/surface/sensor geometry. FRANKLIN (1994) also integrated DEM data with digital satellite data to his study involving the discrimination of sub-alpine forest species and canopy density. He notes that a digital elevation model can be used to significantly increase classi-fication accuracy by adding DEM data to the decision rules.

2.3.3 DEM AND ITS APPLICATION IN THAILAND

Due in part to the importance of DEM data that can be used in a wide variety of applications, the US Geological Survey (USGS) has undertaken the production of 7.5-minute quad-based (1:24,000-scale), high resolution (30-metre grid cell) DEM data for the contiguous 48 states. In addition, the 1:250,000-scale (1-degree) series of DEMs have been also produced by the Defense Mapping Agency and distributed by the USGS, available for the whole of the conterminous United States. In the mean time, while the USGS had produced and distributed the series of DEM data, there has no any government agency in Thailand that has taken a role as a producer or a distributor of such data.

Unfortunately, there are only a few reports published on the use of DEM data in Thailand. One study involving the use of DEM data was reported by HASTINGS et al. (1991). The study was carried out by the Thailand Development Research Institute (TDRI) to introduce GIS technology into natural resources management. In this study, DEM data, in particular slopes, were derived by the terrain modeling function in the TIN™ module of TDRI's GIS. AUNG (1991) also used a DEM interpolated from topographic map for producing slope and aspect maps used for landslide susceptibility mapping in a small part of Nakhon Si Thammarat Province, in southern Thailand.

The UNOCAL Thailand company has generated DEM-like data using x, y, z coordinates in some parts of Thailand, namely the gulf of Thailand, in north-eastern Thailand. Unfortunately, these data are reserved for use in the activity of the company; more information and source of data, therefore, can not be distributed. SILAPATHONG (1992) also integrated topographic data with a GIS into mangrove forest management in the Khlung mangrove forest of Thailand. He produced an altitude map by means of the non-linear interpolation method of the Spatial ANalysis System (SPANS), a microcomputer-based GIS package. Topographic data derived from topographic and hydrographic maps were digitized and used to create elevation data included in a GIS database used in the study.

Furthermore, it was found that a number of researchers tried to include topographic data in their studies. However, they have faced difficulties owing to a lack of good sources for DEM-like data. For example, the JAPAN INTERNATIONAL COOPERATION AGENCY (JICA) (1988) prepared various topographic data used as a basis for site analysis and land use classification in a model area in Thailand. Because of a lack of DEM data manual in situ schemes have mostly been used to produce the topographic data. Most altitude data are presented in the terms of intervals between contour lines. Aspects of the slope falling in each interval are measured in terms of nine azimuths. Gradients are measured in terms of the number of contour lines inside the inscribed circle of each mesh. This indicates that each step of obtaining these data is time consuming and must be performed with care to get reliable results. Actually, it would not be so difficult if a digital elevation model had been adopted instead of such a method. DITBANJONG (1990) included topographic data into one of his studies. This study is relevant to mapping erosion risks using land use, soil, and topographic data. In the study, slope and elevation maps were produced. These topographic data were also derived by a similar method of JICA. CHANGJATURUS (1989) and TOOLPENG (1992) also included non-DEM derived topographic data such as slope and elevation into their studies. SANGUANPONG (1993) built a physical and biological database of Sakaerat Environment Research Station using a GIS. In the study, slope and elevation derived from topographic maps were included into the database. Slope data were produced manually using an interpolation method similar to the one used by JICA. ONGSOMWANG (1993) also tried to classify the relief of his study area using the contour intervals available on topographic maps. This shows that topographic data are essential for many kinds of studies in Thailand. Although topographic data are essential in many studies in Thailand, it can be noticed that the use of topographic data in the form of DEM has been not so widely used according to the lack of data sources and techniques of data production.

Theoretically, a landform is usually perceived as a continuously varying surface that cannot be modeled appropriately on a map. In situ topographic data such as contour lines in topographic maps are, however, used for the display of continually varying surfaces. They are not particularly suitable for numerical analysis or modeling in modern GIS systems. This is why a DEM has become the critical data source in many studies. Hence, the attempt of constructing digital topographic data in the forms of DEM and TIN was carried out in this study. The presentation of using the resulting data in various ways in this study was also concretely done. Resulting terrain data were used to be incorporated into the procedures of the land use and land cover classification of the study.

2.4 Remote Sensing, GIS, and DEMs, tools for land and natural resources management planning in Thailand

Since remote sensing technology has developed rapidly, natural resources survey and monitoring by satellites in Thailand have also been extensively developed. Remote sensing data are considered very important as data source in the use of GIS in natural resources management and for updating current information (TRSC, 1993). Natural resources management is a field where information is crucial, and where information has been acknowledged as the first requirement to successful management of natural resources and environment (CLARK, 1990). As stated before, a DEM is considered an especially important data set in many natural resources spatial databases. Due to the fact that it occurs rapid changes in natural resources and that there is an ongoing attempt to restore and manage them by various government departments as well as non-government organizations, the integration of satellite remote sensing, DEMs, and GIS has become increasingly important in Thailand. This study is one of the attempts that contributes to those tools.

PART I
DATA ANALYSIS
(SATELLITE DATA ANALYSIS FOR LAND USE AND LAND COVER CLASSIFICATION)

3 SELECTION OF REFERENCE AREAS IN THAILAND

3.1 Reasons for the selection

Thailand is situated in the central part of the Southeast Asian mainland and covers approximately an area of 513,115 sq. km. It is located between 5° 45' and 20° 30' latitude and between 97° 30' and 105° 45' longitude. It is divided into four natural regions: (1) the North, (2) the Central Plain, or Chao Phraya River Basin, (3) the Northeast, or the Korat Plateau, and (4) the South, or Southern Peninsula. The North is a mountainous region comprising natural forests, ridges and deep, narrow, alluvial valleys. The Central Plain, the basin of the Chao Phraya River, is a lush, fertile valley. It is the richest and most extensive rice-producing area in the country. The Northeast region, or the Korat Plateau, is a region characterised by a rolling surface and undulating hills. Harsh climatic conditions often occur in this region, especially in the face of floods and droughts. The Southern region is a peninsula along the gulf of Thailand and the Andaman sea. It is hilly to mountainous, with thick virgin, moist, evergreen forests, mangrove forests, and rich deposits of minerals and ores. This region is the center for the production of rubber and the cultivation of other tropical crops.

In accordance with the objectives of the study, four study areas were initially selected from four regions. The principle consideration was the physical characteristics of the areas. Thus, selected areas should include a wide range of topography, geomorphology, vegetation cover types, land use activities, and a fairly wide variation of ecological characteristics.

3.2 Selected study areas

As noted above, four sites were selected for this study as follows:

(1) Ngao Demonstration Forest: a model area in mountainous land, one part of Ngao Demonstration Forest
(2) Angthong Province: a model area in the Central Plain
(3) Amphoe Phu Wiang: a model area in the Korat plateau
(4) Khlung mangrove forest: a model area in the coastal zone.

Since geographic information is the basis for many studies in the field of Remote Sensing, resources and land management, it seems reasonable to include important information describing the study areas, in particular physical characteristics. Thus, an overview of study area characteristics was done based on available and existing data from various sources.

4 GENERAL DESCRIPTION OF STUDY AREAS

The selected study areas are generally different in their main physical characteristics. According to different physical characteristics of areas, these areas also cover various kinds of land use and land cover types. The following descriptions are oriented to some aspects as basic information that can be especially essential for the use in the study. Data are derived from various sources. Most of the data were compiled and reorganized to be presented here.

4.1 Location

Study area I (Mountainous area). This area is located in Ngao Demonstration Forest area that extends over the headwaters of Ngao river in the north-west of Lampang province. As a study area, a 25x25 km square block of the Ngao Demonstration Forest was selected. The study area is located in Ngao district, Lampang province situated in northern Thailand. The area lies between 18° 45' 23.05'' to 18° 59' 00.46'' North, and 99° 45' 36.39'' to 99° 59' 46.62'' East.

Study area II (Central Plain). A 25x25 sq. km block of the middle part of Ang Thong Province was selected as a study area for the Central Plain. The study area lies between 14° 30' 51.54'' to 14° 44' 30.02'' North, and 100° 15' 15.54'' to 100° 29' 06.10'' East. It is composed of a large plain of continuous paddy fields

Study area III (Korat plateau). A 25x25 sq. km block of Khon Kaen province was selected as a study area for the Korat plateau. This area covers Phu Wiang district and a small part of Nong Rua district. The study area is located between 16° 31' 13.05'' to 16° 44' 35.20'' North, and 102° 15' 39.45'' to 102° 29' 53.62'' East.

Study area IV (Coastal zone). Khlung mangrove forest is a part of Chantaburi province situated in the eastern coast of the Gulf of Thailand. As a study area, a 25x25 sq. m along a coastline was selected. This study area lies between 12° 15' 16.35'' to 12° 28' 41.03'' North, and 102° 08' 53.76'' to 102° 22' 49.16'' East.

Figure 1 shows the location diagram of the four study areas.

4.2 Topography and geomorphology

Study area I (Mountainous area). The main topographic features of the Ngao Demonstration Forest are mountainous. This study area is composed of plateau land surrounded by mountains, with the plain areas along the length of the main rivers in the south-eastern part of the area. In this area, two parallel north-south oriented hill ridges are found. The first is located westward from Ngao district and north-south oriented to Muang district. Between these two mountainous ranges, isolated mountains also occur directly below the basin of Ngao district. The elevation of this area varies from 260 m to 1100 m above mean sea level (MSL).

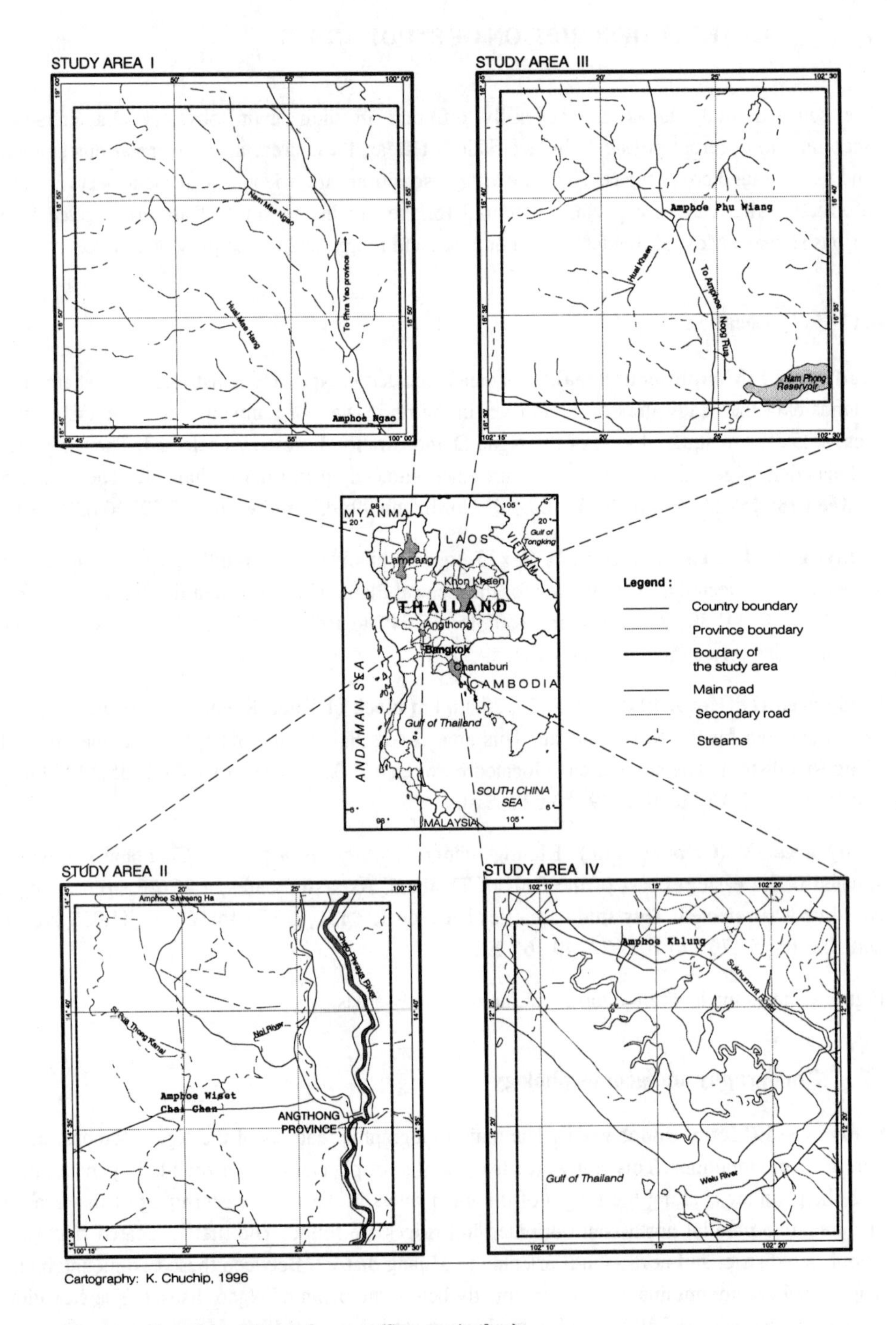

Fig. 1: Location of the study areas (Source: Author).

Based on the classification of the Department of Land Development, Bangkok, in 1987, the geomorphology of the area can be identified into two types as follows:

(1) Semi-recent alluvial terraces. This type was found in plain or relatively plain areas with the slope angle of about 0-2 percent. The area is characterized by alluvial deposits transported through rivers and streams. Soil textures in this area vary from sand to clay.

(2) Old alluvium. This type was found in the old terrace areas and where there are fan-like alluvial deposits. The physiography of this area is undulating to rolling with a gradient of about 1-16 percent. It was developed from alluvial deposits that occur in the Pleistocene period. Transported materials are comprising of gravel, boulder, sand, silt and clay particles. Soils which occur in plain or relatively plain areas have not so good permeability. Soils which occur in relatively rolling areas have better permeability.

Study area II (Central plain). Unlike the study area just described above, the topographic features of this study area is mostly plain with a bowl-like land surface formation. There are many small canals and swamps all over the area. The Chao Phraya river is the most important river in this region. The altitude varies slightly from 2 to 9 meters above mean sea level (MSL). This study area is mainly a flood plain where there are floods during the rainy season almost every year. Some parts of the area consist of low terraces. These terraces are composed of alluvial deposits. These areas are relatively plain to slightly rolling. Slopes of these terrace areas vary from 0.1% to 2.0%.

Study area III (Korat plateau). The study area is an interesting area covering a part of the upper watershed of the Nam Pong reservoir, a large multipurpose reservoir-irrigation system in Northeast of Thailand. The north-western portion of the area is located an important valley, called Phu Wiang. It is relatively complex with a continuous succession of different slopes. These slopes are especially steep in the exterior. The altitude of the area varies from 180 to 500 meters above MSL.

According to the type of parent materials, relief, surface configurations, and drainage patterns, this area can be differentiated into five main landform units as follows:

(1) Alluvial flood-plain. It consists characteristically of sandy loamy soils that occur in narrow flat strips along both sides of the stream bed. This plain is often flooded during the rainy season.

(2) Low alluvial plain. It covers most of the study area. This plain alternates with the nearly flat to middle terrain, and rises gradually above the alluvial flood-plains. This terrain is mostly used for rice fields. The soil in this area is characterized by sandy loam on the surface and finer texture material in the subsoil.

(3) Middle undulating terrain. This area is composed of undulating slopes scattered throughout the study area. Most of the soil in this terrain consists of sand sediments with slightly less clay

in the surface layers. Most of the soil is used for upland crops such as sugar cane, corn, cassava, and kenaf.

(4) Erosional surface. It mostly occurs in the surrounding areas of the Central Plain. The topography is undulating to gently rolling, some areas are found on low flat hills, irregularly distributed throughout the flat land of the low terrain. Most of the soils on this terrain are sandy loam or sandy clay loam. The surface texture is underlined by a clay loam, silty clay, or clay subsoil.

(5) Slope complex. It occurs in the north part of the study area. A hill ridge is located to the north and slightly westward to the south.

In association with geomorphology of the area, many parts of the area are mountains covering with natural forests. Agricultural areas are usually found in alluvium plains.

Study area IV (Coastal zone). The topographic features of this zone can be classified based on the study of the Department of Land Development, Bangkok, in 1983, into 5 categories as follows:

(1) Beach zone. This zone lies along the Gulf of Thailand 1 to 3 meters above MSL.

(2) Inter-tidal zone. This zone consists of plain areas that are situated usually between the beach and plain lands. These areas are affected from tides of the sea. The tides are diurnal, with an average amplitude of 2.5 meters for spring tides and 0.6 meters for neap tides above the lowest low waters (SILAPATHONG, 1992). Formerly, all of the areas have been covered with wealthy continuous mangrove forest, named Khlung mangrove forest. Unfortunately, this mangrove forest has been severely depleted since the past decade due to illegal cutting.

(3) Plain areas. These areas lie parallel to the coastline of the gulf. The altitude of the areas varies from 3 to 5 meters above MSL. Some parts of the area consist of scattered small hills ranging in height between 140-210 meters.

(4) Rolling to steep rolling areas. This type occurs in the east and north-east of the study area. The altitude of the areas varies from 30 to 150 meter (MSL).

(5) Mountainous areas. These areas occur in the north and south-east of the study area. The altitude of these mountains varies from 300 to 1,670 meters. Most of these areas covered with moist evergreen forests that are still in a good condition.

4.3 Geology

Study area I (Mountainous area). Based on the geological maps at a scale of 1:250,000 prepared by the Department of Mineral Resources, Bangkok, in 1971, the rocks of the study area in Ngao Demonstration Forest consist of the following sequences: (The geological map of this area is shown in Fig. 2.)

(1) Quaternary Rock (Qa). This geological unit belongs to the Alluvium Group. Unconsolidated rocks of the Quaternary are river gravel, sand and mud. It is considered to occur in the recent age of the Quaternary. This type of material found in this study area covers an area of 2,238.375 hectares, or 3.58 percent of the total area.

(2) Higher Terrace Quaternary (Qt 2). This geological unit is included in the Mae Teng Group. Rocks of the higher terraces include gravel, sand, silt and clay. It is considered to be formed in the Pleistocene age of the Quaternary period. The Higher Terrace Quaternary found in this study area covers an area of 700.562 hectares, or 1.12 percent of the total area.

(3) Tertiary Rock (T). This geological unit belongs to Mae Mo Group. Its rocks consist of fresh water sandstone, shale, carbonaceous shale, limestone, viviparous beds, and lignite. It occurs in the Pliocene/Miocene age of the Tertiary period. The Tertiary rocks found in this study area cover an area of 8,245.875 hectares, or 13.19 percent of the total area.

(4) Phu Kradung Formation of Middle Jurassic Age (JR 2). This formation belongs to the Korat Group that is considered to be developed in the middle Jurassic. Rocks found in this unit are shale, siltstone, fine grained sandstone with a reddish-brown color, ripple marked, and cross-bedded. The Jurassic rocks of Phu Kradung Formation found in this study area cover an area of 1,779.437 hectares, or 2.85 percent of the total area.

(5) Nong Hoi Formation of Triassic Age (TR 3). This geological unit belongs to the Lampang Group that is considered to belong to the Carnian to Ladinian age of the Mesoic era. Rocks of this unit include greenish-gray shale, sandstone, tuffaceous sandstone, laminated shales, and conglomerates. The fossils of Halobia, Daonella, Posidonia, Trachycera, Paratrachycera, Joannite, can be found in this rock type. These Triassic rocks of Pha Daeng Formation found in this study area cover an area of 21,876.562 hectares, or 35 percent of the total area.

(6) Phra Kan Formation of Triassic Age (Tr 2). This geological unit is also included in the Lampang Group that is considered to belong to the Ladinian to Anisian Age of the Triassic period of the Mesozoic era. Rocks of this unit are limestones with massive or banded, dark gray to medium gray color; calcareous shale, sandstone, with gray to greyish-brown colour and well stratified. The fossils of brachiopods, Claraia, Halobia, Daonella, Posidonia, and ammonites can be found in this rock type. The Triassic rocks found in this study area cover an area of 214.312 hectares, or 0.34 percent of the total area.

(7) Huai Thak Formation of Permian Age (Pm3). This geological unit belongs to the Ratburi Group that is considered to belong to the Kazanian to Kungurian age of the Permian period. Rocks in this unit comprise of shale, carbonaceous shale, calcareous shale, tuffaceous shale and sandstone, and laminated shale with fossils of Dielasma, Leptodus, Orthotichia, Echinochus, Neospirifera, Schizophoria, Aviculopecten, etc. The Permian rocks found in this study area cover an area of 25,731.812 hectares, or 41.17 percent of the total area.

(8) Pha Huat Formation of Permain Age (Pm2). This geological formation also belongs to the Ratburi Group that is considered to be found in the Artinskian to Sakmarian age of the Permain

period. Rocks in this unit are massive limestone, shale, calcareous shale, laminated shale and tuffaceous sandstone, tuff, chert nodules, and chert beds with fossils of fusulinids, Fenestella, Agatheceras, Bellerophon, corals, etc. These rocks found in this study area cover an area of 1,713.062 hectares, or 2.74 percent of the total area.

Figure 2 shows the distribution of geological units in the study area.

Study area II (Central Plain). Based on the geological maps at a scale of 1:250,000 prepared by the Department of Mineral Resources, Bangkok, in 1976, the geological units of this study area are composed of holocene to recent unconsolidated rocks. Figure 3 shows the distribution of geological units in the study area. There are two geological units found in this study area, namely:

(1) Recent flood plain alluvials (Q). This area consists of sands, silts, and swamps of holocene to recent development. In this study area, this type cover an area of 59,101.687 hectares, or 94.56 percent of the total area.

(2) Old alluvial fans (Q1). This area consists of colluvial and old flood plain deposits of high and low terraces which consist of gravel, sand, silt, and laterite. This type is considered to have been developed in the Pleistocene. This type covers an area of 3,398.312 hectares, or 5.44 percent of the total area.

Study area III (Korat plateau). Based on the geological maps at a scale of 1:250,000 prepared by the Department of Mineral Resources (1979), the rocks in this area mainly belong to the Korat Group. The lithological group consists of various formations that can be identified as follows:

(1) Alluvial deposits (Qa). This type developed during the Quaternary Age. It consists of gravel, sand, silt, and clay. In the study area, this geological type covers an area of 9,787.625 hectares, or 15.66 percent of the total area.

(2) Khok Kruat Formation (Kkk). This formation belongs to the Korat Group and is considered to have been formed during the Upper Cretaceous. It consists of sandstone, siltstone, shale, and lime-noduled conglomerate that have a reddish-brown, gray, grayish-white, and brown colour with gypsum at the upper part. This geological type covers an area of 988.625 hectares, or 1.58 percent of the total area.

(3) Phu Phan Formation (Kpp). This formation also belongs to the Korat Group and is considered to have been formed during the Lower-Middle Cretaceous Age. It consists of white or pale-orange sandstone that commonly mixes with pebbles of up to 5 cm in diameter. The pebbles usually consist of quartz, chert, red siltstone, and igneous rocks. Some of them are found cross-bedded shale or as inter-bedded conglomerate. This type underlies some plains and has low to medium permeability. This geological type covers an area of 1,698.5 hectares, or 2.72 percent of the total area.

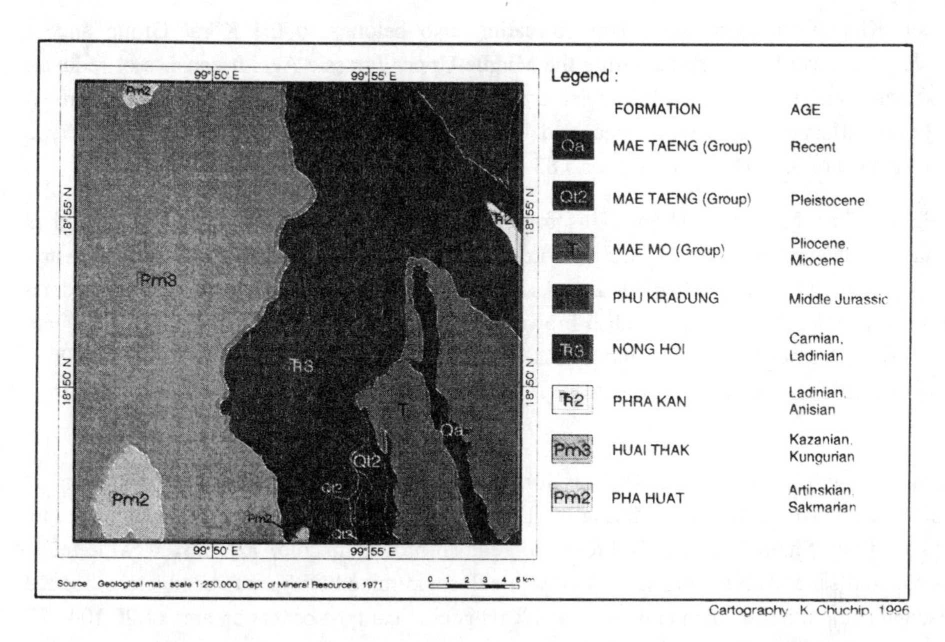

Fig. 2: **Geological map of the study area I in the mountainous area (Source: Prepared by the author, based on the Geological map (Scale 1:100,000) of the Department of Mineral Resources, 1971).**

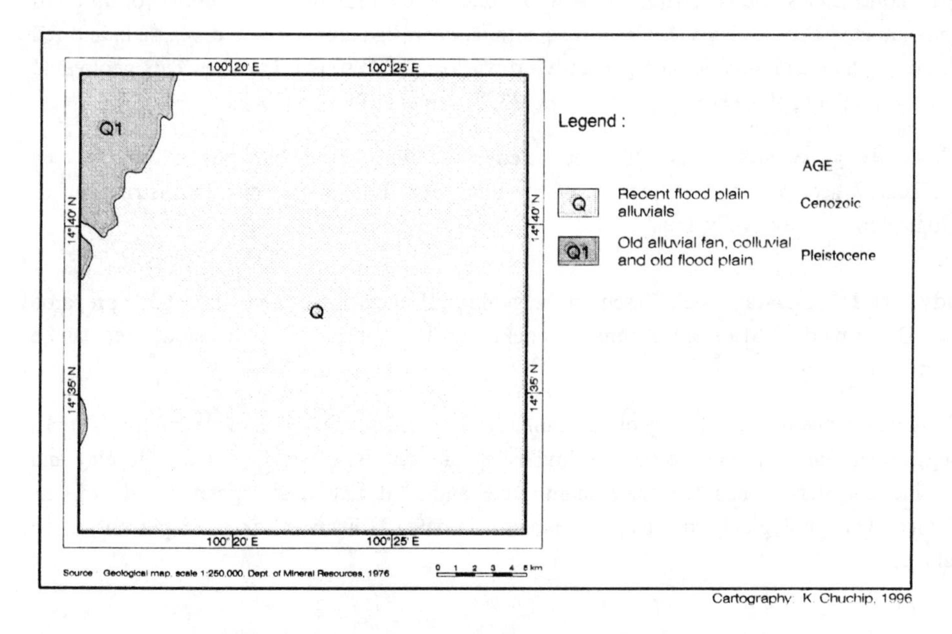

Fig. 3: **Geological map of the study area II in the Central Plain (Source: Prepared by the author, based on the Geological map (Scale 1:100,000) of the Department of Mineral Resources, 1976).**

(4) Sao Khua Formation (Jsk). This formation also belongs to the Korat Group and is considered to have been formed during the Middle-Upper Jurassic Age. It consists of reddish-brown sandstone that mixes with grayish-brown and reddish-brown siltstone, purplish-brown and brick-red shale mixed with mica, and lime-noduled conglomerate. This geological type covers an area of 3,640.812 hectares, or 5.83 percent of the total area.

(5) Phra Wihan Formation (Jpw). This formation also belongs to the Korat Group and is considered to have been formed during the Lower-Middle Jurassic Age. It consists of white and pink sandstone. In the upper bed, these rocks are generally massive and consist of cross-bedded layers of pebbles with some reddish-brown and gray shale and conglomerate. This type underlies the surrounding mountainous area and has a higher permeability than the Phu Phan formation. This geological type covers an area of 3,750.875 hectares, or 6.0 percent of the total area.

(6) Phu Kradung Formation (Jpk). This formation also belongs to the Korat Group and is considered to have been formed during the Lower Jurassic Age. It consists of shale, siltstone, and sandstone. Shale is generally brown to reddish-brown in color and mixed with mica. Siltstone and sandstone are brown and gray in color and mixed with small scale cross-bedded mica, and some lime-noduled conglomerate. This geological type covers an area of 20,104.937 hectares, or 32.17 percent of the total area.

(7) Nam Phong Formation (TRnp). This formation also belongs to the Korat Group and is considered to have been formed during the Upper Triassic Age. It consists of sandstone, shale, and siltstone. Sandstone is generally brown to reddish-brown in color with pebbles of up to 10 cm in diameter. Pebbles consist of quartz, quartzite, chert, igneous rocks, red siltstone, and red sandstone. Shale and siltstone are generally brown to reddish-brown in color. This geological type covers an area of 20,808.312 hectares, or 33.29 percent of the total area.

Additionally, some parts of this study area are covered with big water surfaces covering an area of 1,720.312 hectares, or 2.75 percent of the total area. Figure 4 shows the distribution of geological units in the study area.

Study area IV (Coastal zone). Based on the geological maps at a scale of 1:250,000 prepared by the Department of Mineral Resources, Bangkok, in 1976, the rocks of this study area can be identified as follows:

(1) Alluvial deposits (Qa). This geological unit consists of sedimentary and metamorphic rocks being formed during the Quaternary. Alluvial deposits consist of sand, gravel, silt, clay and estuarine mud-flats. Beach deposits contain gravel and sand. Lagoonal deposits consist of silts and mud. This geological type covers an area of 33,049.187 hectares, or 52.88 percent of the total area.

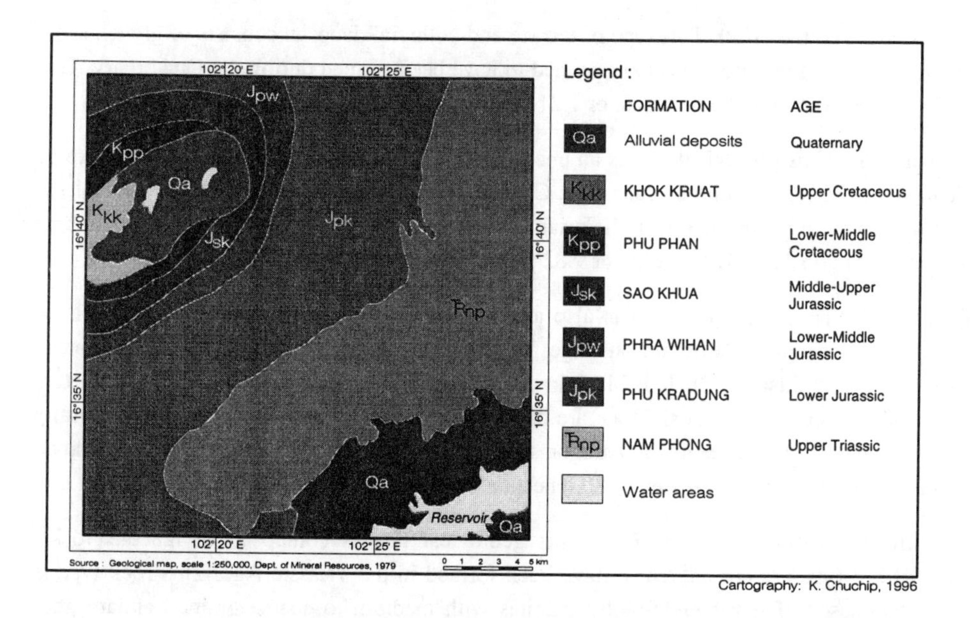

Fig. 4: **Geological map of the study area III in the Korat Plateau (Source: Prepared by the author, based on the geological map (Scale 1:100,000) of the Department of Mineral Resources, 1979).**

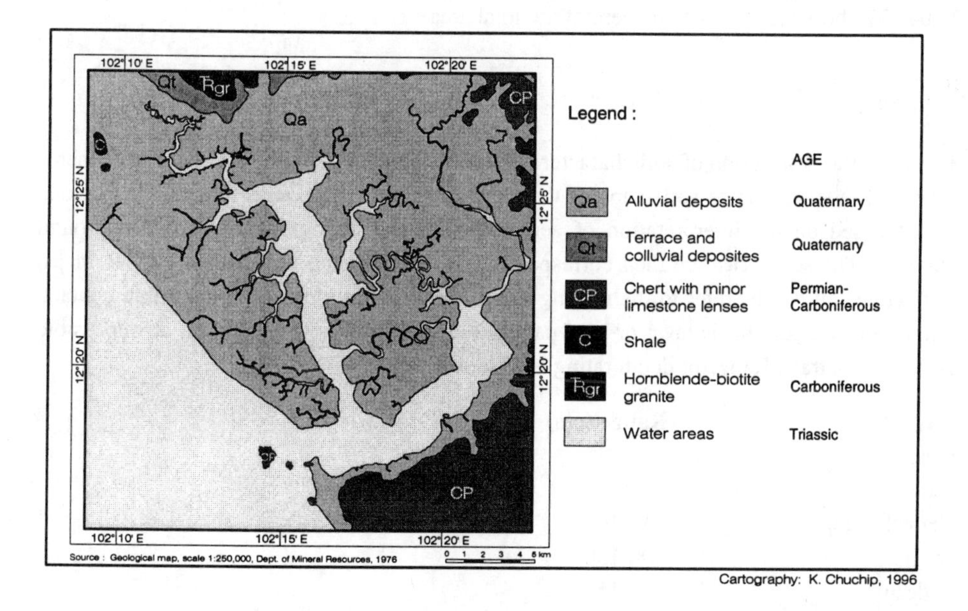

Fig. 5: **Geological map of the study area IV in the Coastal Zone (Source: Prepared by the author, based on the geological map (Scale 1:100,000) of the Department of Mineral Resources, 1976).**

(2) Quaternary Terrace(Qt). This type of terrace and colluvial deposits consists of gravel, sand, silt, clay, mud, and laterite. It is also assumed to have been formed during the Quaternary. This type covers an area of 757.5 hectares, or 1.21 percent of the total area of the study area.

(3) Chert (CP). This geological unit is an unconformity. Rocks of this type are considered to be formed in the Permian-Carboniferous Age. Chert rocks found in this study area are gray, black, and red in color with well-bedded radiolarian, and minor limestone lenses. This geological type covers an area of 6,139.937 hectares, or 9.82 percent of the total area of the study area.

(4) Shale (C). This geological unit is also an unconformity. This rock is considered to have been formed during the Carboniferous Age. This rock is mainly composed of shales, or inter-bedded shale and siltstone. Shale is black in color with carbon layers (approximately 50 meters thick). Inter-bedded shale and siltstone are dark brown to black in color. They occur generally with well-bedded carbonaceous and minor sequences of conglomeratic sandstone, sandstone and shale. This type covers an area of 80.0 hectares, or 0.13 percent of the total area.

(5) Hornblende-biotic granite (TRgr). This geological unit is composed of igneous rocks. Rocks of this type are considered to have been formed in the Triassic Age. This rock type is mainly composed of hornblende-biotic granites with medium to coarse grained andamygda-loidal. This type covers an area of 432.0 hectares, or 0.69 percent of the total area.

Figure 5 shows the distribution of geological units in the study area.

Additionally, some parts of this study area cover with big surfaces of water that cover an area of 22,041.375 hectares, or 35.27 percent of the total area.

4.4 Soils

In this study, the description of soil characteristics of the study areas is based on the documents and detailed reconnaissance soil maps of the Department of Land Development. These soils maps are based on the interpretation of aerial photographs, topographic maps, and ground survey data. The soil series are then corresponding to the mapping unit name of those maps. The effective soil depths refer to the rooting zone where the limiting depth is a lithic contact, paralithic contact, petroferric layer or hard pan, through which it is very difficult or impossible for roots to penetrate. Range of depth rating is as follow:

Rating	**Range** (cm)
Very shallow	<25
Shallow	25-50
Moderately deep	50-100
Deep	100-150
Very deep	>150

The drainage of soil is based on the ratings described in the USDA Soil Survey Manual. The permeability is based on field observations of the soil profile; least permeable horizon of the solum or immediate substratum determines permeability of the soil. Definition of ratings is as follows:

Slow: Soils expected to have hydraulic conductivity of less than 0.5 cm/hour
Moderate: Soils expected to have hydraulic conductivity of 0.5 to 15 cm/hour
Rapid: Soils expected to have hydraulic conductivity more than 15 cm/hour

The surface runoff is estimated, based on characteristics of the soil profile, soil slope, climate and vegetation cover. Definition of ratings is as follows:

Slow: Surface water flows away so very slowly that free water lies on the surface for the considerable periods of time or immediately enters the soil. Much of the water either passes through the soil or is lost to evaporation. Soils with slow runoff are subject to little or no erosion hazard.

Medium: Surface water flows away at such a rate that a moderate amount of water enters the soil profile and free water lies on the soil surface for only short periods. Most of the precipitation is absorbed by ground channels. With medium runoff the loss of water over the surface does not seriously reduce the supply available for plant growth. Erosion hazards can be expected to be slight or moderate if such soils are cultivated.

Rapid: A large or very large proportion of the precipitation moves rapidly over the surface of the soil and very little moves through the soil profile. Surface water moves as fast or almost or fast off the soil as it is added to the soil. Erosion hazard is moderate, high or very high.

The period of water saturation indicates the length of time that the soil surface and/or subsurface is at or above field capacity. Saturation by rainwater, seepage, river water or sea water; but *not* by irrigation water.

Study area I (Mountainous area). Based on the classification of the Department of Land Development, Bangkok, in 1981, soil types found in this study area can be identified as follows:

(1) Sanphaya series (Sa). This soil series is a recent alluvium that is generally found in flood plains. It can sometimes be found as a narrow strip near a levee. Its physiography is flat or relatively flat with a gradient of 0-2 percent. It is a very deep soil and has a moderate to well drainage. The water holding capability is moderate. The soil type found in the study area covers an area of 181.187 hectares, or 0.29 percent of the total area.

(2) Tha Muang series (Tm). The soil parent material of this series also belongs to the recent alluvium. The soil physiography is relatively flat and moderately well drained. Surface runoff on this soil series is medium. The soil texture varies from sandy loam to loam over silty clay.

This soil type found in the study area covers an area of 67.250 hectares, or 0.11 percent of the total area.

(3) Alluvial soil poorly drained (As-p). This series is also part of the recent alluvium and formed on river levees. The soil physiography is flat in the narrow valleys. Its gradient is about 0-1 percent. The soil capacity for water holding is moderate to high. Surface runoff in this area is slow to moderate. This soil type found in the study area covers an area of 60.937 hectares, or 0.10 percent of the total area.

(4) Alluvial complex (Ac). This series is recently developed and generally formed on river levees. Parent material is recent flood plain alluvium. The soil physiography is flat to slightly flat with its gradient less than 2 percent. Soil capacity for water holding is moderate to high. Surface runoff in this area is slow to moderate. This soil type found in the study area covers an area of 364.625 hectares, or 0.58 percent of the total area.

(5) Mae Sai series (Ms). This series belongs to a sub-recent terrace group. Parent material is a sub-recent to recent alluvium. It is developed and formed on river levees. The soil texture consists of sandy loam with silt, sandy loam over silty clay, or clay. Soil drainage is relatively poor. This soil type found in the study area covers an area of 1,293.473 hectares, or 2.07 percent of the total area.

(6) Hang Dong series (Hd). This series also belongs to a sub-recent to recent terrace. Parent material is a semi-recent alluvium. Its physiography is gently flat with a gradient of 0-2 percent. It is a deep soil and poorly drained. Soil capability for water holding is high. This soil type found in the study area covers an area of 180.5 hectares, or 0.29 percent of the total area.

(7) Phan series (Ph). This series belongs to a sub-recent to recent terrace group. Parent material is a sub-recent alluvium. Its physiography is gently flat with a gradient 0-2 percent. It is a very deep soil and poorly drained. Soil capability for water holding is rather high. This soil type found in the study area covers an area of 79.062 hectares, or 0.13 percent of the total area.

(8) Mae Sai/ Hang Dong association (Ms/Hd). This association is composed of the Mae Sai series and the Hang Dong series with the ratio of 60 and 40 percent, respectively. Generally, both series are formed continuously or alternately together in the same area. This soil type found in the study area covers an area of 61.312 hectares, or 0.10 percent of the total area.

(9) Kamphaeng Saen series (Ks). This series is a semi-recent alluvium that formed on old levees. Its physiography is flat to gently undulating with a gradient of 1-3 percent. It is a deep and well drained soil. The soil permeability is moderate. This soil type found in the study area covers an area of 102.375 hectares, or 0.16 percent of the total area.

(10) Lampang series (Lp). This series can be found on areas where old alluvium and fan-liked deposits occur. The soil parent material is old alluvium. The soil physiography is flat to nearly undulating with a gradient of 0-3 percent. It is a deep soil with poor drainage. This soil type found in the study area covers an area of 24.437 hectares, or 0.04 percent of the total area.

(11) San Pa Tong series (Sp). This series is formed on low to middle terraces. The soil physiography is nearly flat to gently undulating with a gradient of 2-7 percent. It is a deep soil with rather good drainage. Soil parent materials are old alluvium. Soil textures are loam, or sandy loam, or clay loam. This soil type found in the study area covers an area of 37.750 hectares, or 0.06 percent of the total area.

(12) Hang Chat series (Hc). This series can be found on areas where terraces and fan-like deposits occur. The soil physiography is gently undulating to undulating with a gradient of 3-12 percent. It is a deep soil with good drainage. Soil parent materials are old alluvium. Soil textures are loam, or sandy loam, or clayey loam. This soil type found in the study area covers an area of 603.562 hectares, or 0.97 percent of the total area.

(13) Mae Rim series (Mr). This series is generally formed on middle to high terraces. Soil parent materials are old alluvium. The soil physiography is gently to steeply undulating with a gradient of 3-16 percent. It is a deep soil with good drainage. Soil textures are loam, or sandy loam, or clayey loam. This soil type found in the study area covers an area of 2,428 hectares, or 3.88 percent of the total area.

(14) Mae Rim/Hang Chat/Satuk association (Mr/Hc/Suk). This association consists of fan-liked deposits. All the series are formed continuously or alternately together in the same area. The Maerim series is mixed with gravel. There are no gravel found in the Hang Chat and the Satuk series. This soil association found in the study area covers an area of 2,074.250 hectares, or 3.32 percent of the total area.

(15) Li series (Li). This series is generally formed on eroded terraces. Soil parent materials are alluvium, colluvium and residuum that are derived from shale and phyllite. It is a shallow and well drained soil. The soil physiography is rolling with a gradient of 7-35 percent. Surface runoff of this soil type is high. This soil type found in the study area covers an area of 135.375 hectares, or 0.22 percent of the total area.

(16) Ban Chong series. (Bg). This series is also formed on eroded terraces. Soil parent materials are colluvium and residuum that are derived from shale and metamorphic equivalent rock. The soil is shallow to medium in depth with good drainage. The soil physiography is flat to hilly with a gradient of less than 35 percent. Surface runoff risk of this soil type is high. This soil type found in the study area covers an area of 169.312 hectares, or 0.27 percent of the total area.

(17) Ban Chong/Muak Lek/Li association (Bg/Kl/Li). This association is composed of the Ban Chong, Muak Lek, and Li series with the ratio of about 50, 25, 25 percent, respectively. They are generally formed from the same parent materials and located continuously or alternately together in the same topographic area. They are grouped together as an association. This soil type found in the study area covers an area of 198.0 hectares, or 0.32 percent of the total area.

(18) Slope Complex (SC). This is a definition of mountainous areas with a gradient of more than 35 percent. Most of the area covers with natural forests. The Department of Land

Development has not identified the soil type of these areas because these areas should be reserved as permanent forest areas which are unsuitable for agricultural purposes. Slope complex found in this study area covers an area of 54,002.25 hectares, or 86.4 percent of the total area.

(19) Rocky Land and Stony Land (Rl and Sl). These areas are composed of rock outcrops. Most of them are limestone rocks from which brown and reddish brown soils were developed. This type found in the study area covers an area of 436.375 hectares, or 0.70 percent of the total area.

Figure 6 shows the distribution of soil units found in the study area.

Study area II (Central plain). Based on the classification of the Department of Land Development, Bangkok, in 1976, soil types found in this study area can be classified as follows:

(1) Bang Len series (Bl). This series is an alluvial soil formed on former tidal flats with a gradient of about 0-1 percent. Soil parent materials are derived from marine and brackish water deposits. Effective soil depth is very deep. Soil texture is clay throughout. Period of water saturation is 6-7 months on the surface, 8-10 months below the subsurface, and for 1-3 months in ground water table below 120 cm. This soil series is poorly drained with a slow permeability and a low surface runoff. This soil type found in the study area covers an area of 1,400.250 hectares, or 2.24 percent of the total area.

(2) Bang Khen series (Bn). This series is also an alluvial soil that was formed on former tidal flats with a gradient of about 0-1 percent. Soil parent materials are derived from marine and blackish water deposits. Effective soil depth is very deep. Soil texture is clay or silty clay over clay. The period of water saturation is 5-6 months on the surface, 8-10 months below the subsurface, and for 1-3 months in the ground water below 120 cm. This soil series is poorly drained with a slow permeability and a slow surface runoff. This soil type found in the study area covers an area of 95.375 hectares, or 0.15 percent of the total area.

(3) Bang Khen series, undulating phase (Bn-u). This series is the Bang Khen series in undulating terrain with a gradient of about 1-2 percent. Other characteristics are quite the same as the Bang Khen series. This soil type found in the study area covers an area of 465.812 hectares, or 0.75 percent of the total area.

(4) Tha Muang series (Tm). This series is a recent alluvial soil that was formed on flood plains with a gradient of about 1-2 percent. Effective soil depth is very deep. The soil profile is stratified, consisting of clay loam and sandy clay loam, or consisting of clay loam, sandy loam, and silty loam. Period of water saturation is 1-2 months on surface, and 1-3 months below the subsurface. This soil series is moderately well drained with a moderate permeability and a medium surface runoff. This soil type found in the study area covers an area of 5,372.625 hectares, or 8.60 percent of the total area.

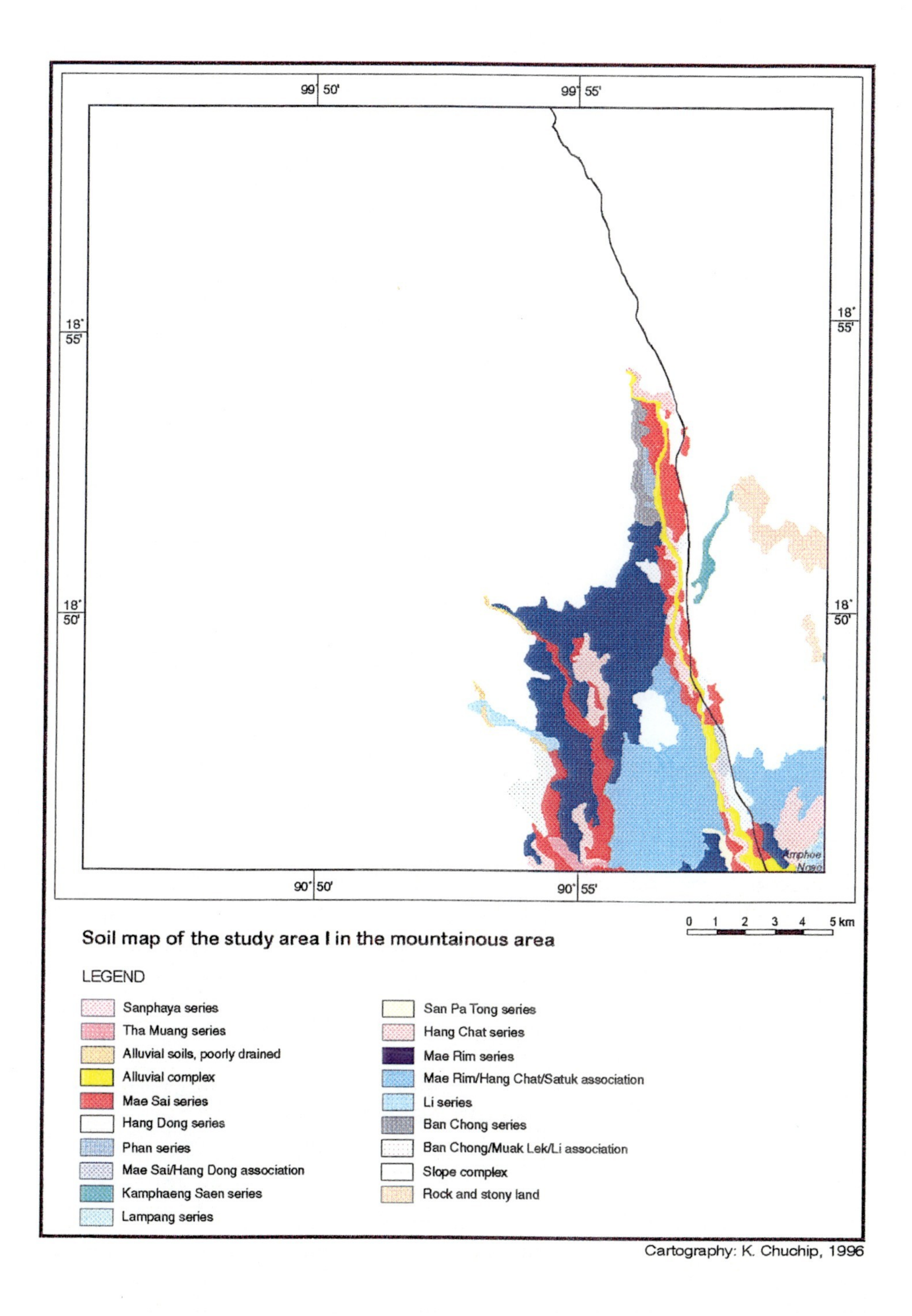

Fig. 6: Distribution of soil units of the study area I in the mountainous area (Source: Prepared by the author, based on the reconnaissance soil map scale to 1:100,000 of the Department of Land Development, 1981).

(5) Sanphaya series (Sa). This series is also a recent alluvial soil that was formed on flood plains with a gradient of about 0-1 percent. Effective soil depth is very deep. The soil profile consists of loam or silty, clay loam over a stratified clay loam or silty, clay loam. Period of water saturation is 4-5 months on the surface, and 6-8 months below the subsurface. This soil series is moderately well drained with a moderate permeability and a slow surface runoff. This soil type found in the study area covers an area of 1,639.250 hectares, or 2.62 percent of the total area.

(6) Chai Nat series (Cn). This series is also a recent alluvial soil that was formed on flood plains with a gradient of about 0-1 percent. Effective soil depth is very deep. The soil profile consists of loam, silty, clay loam, or clayey loam over silty, clay loam or silty clay. Period of water saturation is 3-4 months on the surface, and 6-8 months below the subsurface. This soil series is somewhat poorly drained with a moderate permeability and a slow surface runoff. This soil type found in the study area covers an area of 6,033.875 hectares, or 9.65 percent of the total area.

(7) Ratchaburi series (Rb). This series is also a recent alluvial soil that was formed on flood plains with a gradient of about 0-1 percent. Effective soil depth is very deep. The soil profile consists of clay, or silty, clay loam over clay or silty clay. Period of water saturation is 4-5 months on the surface, and 6-8 months below the subsurface. This soil series is somewhat poorly drained with a low permeability and a slow surface runoff. This soil type found in the study area covers an area of 13,467.875 hectares, or 21.55 percent of the total area.

(8) Ratchaburi series, undulating phase (Rb-u). This series is the Ratchaburi series in undulating terrain with a gradient of about 1-2 percent. Other characteristics are similar to the Ratchaburi series. This soil type found in the study area covers an area of 39.437 hectares, or 0.06 percent of the total area.

(9) Ratchaburi/Singburi association (Rb/Sin). This series is an association of the Ratchaburi and the Singburi series. They are generally formed from the same parent materials and located continuously or alternately together in the same topographic area. According to the scale of the map used, these series cannot be easily differentiated. They are grouped together as an association. Soil characteristics depend on each associated soil series. This soil type found in the study area covers an area of 96.250 hectares, or 0.15 percent of the total area.

(10) Singburi series (Sin). This series is also a recent alluvial soil which was formed on flood plains with a gradient of about 0-1 percent. Effective soil depth is very deep. Soil texture is clay throughout. The period of water saturation is 6-7 months on the surface, 10-11 months below the subsurface. This soil series is poorly drained with a slow permeability and a slow surface runoff. This soil type found in the study area covers an area of 20,960.312 hectares, or 33.54 percent of the total area.

(11) Singburi series with light textures subsoil (Sin-l). Soil texture is clay over sandy, clay loam or loam. The period of water saturation is 6-7 months on the surface, and 10-11 months below the subsurface. This soil series is poorly drained with a moderate permeability and a slow

surface runoff. This soil type found in the study area covers an area of 172.187 hectares, or 0.28 percent of the total area.

(12) Singburi series, undulating phase (Sin-u). This soil type is the Singburi series in undulating terrain with a gradient of about 1-2 percent. Other characteristics are similar to Singburi series (Sin). This soil type found in the study area covers an area of 172.125 hectares, or 0.28 percent of the total area.

(13) Kamphaeng Saen series (Ks). This series is a sub-recent alluvial soil that was formed on low terraces with a gradient of about 1-2 percent. Effective soil depth is very deep. Soil texture is loam, clay loam or sandy, clay loam over clay loam or silty, clay loam. Groundwater is below 100 cm the year round. This soil series is well drained with a moderate permeability and a slow surface runoff. This soil type found in the study area covers an area of 1,095.0 hectares, or 1.75 percent of the total area.

(14) Nakorn Pathom series (Np). This series is also a sub-recent alluvial soil that was formed on low terraces with a gradient of about 0-1 percent. Effective soil depth is very deep. Soil texture is clayey loam or silty, clayey loam over clay or silty clay. Period of water saturation is 4-5 months on the surface, and 8-10 months below the subsurface. This soil series is somewhat poorly drained with a slow permeability and a slow surface runoff. This soil type found in the study area covers an area of 7,100.250 hectares, or 11.36 percent of the total area.

(15) Saraburi series (Sb). This series is also a sub-recent alluvial soil that was formed on low terraces with a gradient of about 0-1 percent. Effective soil depth is very deep. Soil texture is clay or silty clay over clay. Period of water saturation is 5-6 months on the surface, and 10-11 months below the subsurface. This soil series is somewhat poorly drained with a slow permeability and a low surface runoff. This soil type found in the study area covers an area of 1,440.625 hectares, or 2.31 percent of the total area.

(16) Saraburi series, high phase (Sb-h). This series is the Saraburi series in high phase. Other characteristics are similar to the Saraburi series. This soil type found in the study area covers an area of 1,892.875 hectares, or 3.03 percent of the total area.

(17) Saraburi series, sandy clay type (Sb-sc). This series is the Saraburi series with a texture of sandy clay over clay. Other characteristics are similar to the Saraburi series (Sb). This soil type found in the study area covers an area of 55.125 hectares, or 0.09 percent of the total area.

(18) Saraburi/Singburi association (Sb/Sin). This series is an association of the Saraburi and the Singburi series. They are generally formed from the same parent materials and located continuously or alternately together in the same topographic area. According to the scale of the map used, these series cannot be easily differentiated. They are grouped together as an association. Soil characteristics depend on associated soil series. This soil type found in the study area covers an area of 666.375 hectares, or 1.07 percent of the total area.

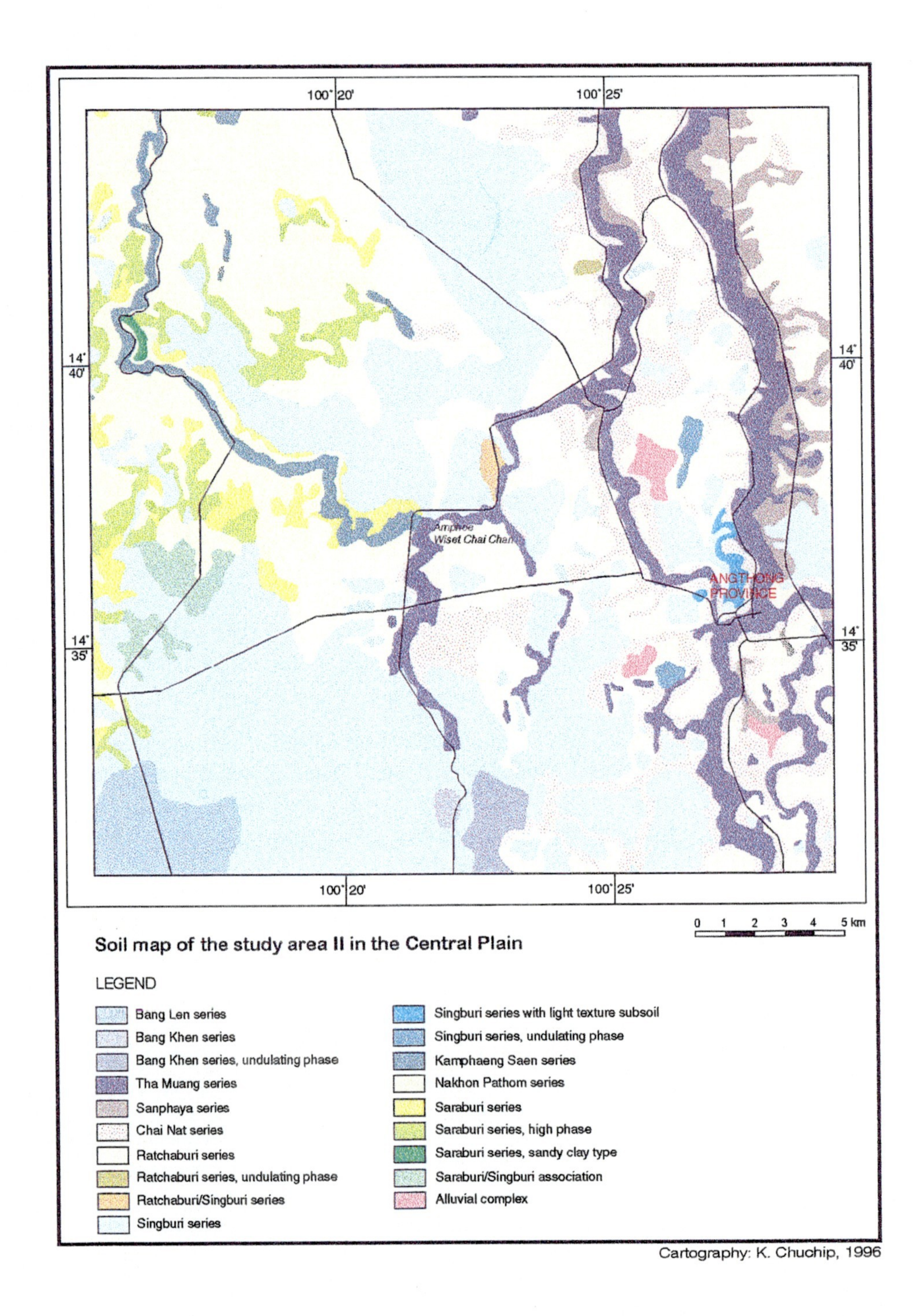

Fig. 7: **Distribution of soil units of the study area II in the Central Plain (Source: Prepared by the author, based on the reconnaissance soil map scale to 1:100,000 of the Department of Land Development, 1976).**

(19) Alluvial Complex (AC). Soil characteristics are the same as described before. This soil type found in the study area covers an area of 334.375 hectares, or 0.54 percent of the total area.

Figure 7 shows the distribution of soil units found in the study area.

Study area III (Korat plateau). Soil types of this area mainly consist of low humic gley, Gray Podozols, Red Yellow Podozols and Regosols. They are mainly derived from sandstone parent materials. Based on the classification of the Department of Land Development, Bangkok, in 1973, soil series found in this study area can be identified as follows:

(1) Alluvial Complex (AC). The parent materials of this soil type are recent and sub-recent riverine alluvium. It is normally formed on flood plains and valley flat. This soil type found in the study area covers an area of 367.737 hectares, or 0.59 percent of the total area.

(2) Si Thon series (St). This series is also formed on flood plains and valley flats with less than 1 percent slope. The effective soil depth is very deep. The soil texture is sandy loam or loamy sand over stratified clayey and sandy materials. The surface is always flooded by over flown stream water and by impounded rain water up to 50 cm depth for 1-2 months during the rainy season. The groundwater table during the peak of dry season is below 1.5 m depth. This soil series is poorly drained with a moderate permeability and a slow surface runoff. This soil type found in the study area covers an area of 82.687 hectares, or 0.13 percent of the total area.

(3) Roi Et series (Re). This series is an old riverine alluvial soil that was formed on low terraces with a range of slope less than 2 percent. The effective soil depth is very deep. The soil texture is sandy loam or loamy sand over loam or sandy clay loam or clayey loam grading to clay. The surface is flooded by impounded rain water up to 30 cm deep for 3-4 months. The groundwater table during the peak of dry season is below 3 m depth. This soil series is poorly drained with a moderate permeability and a slow surface runoff. This soil type found in the study area covers an area of 13,833.937 hectares, or 22.13 percent of the total area.

(4) Roi Et series, loamy variant (Re-l). This series is also an old riverine alluvial soil that was formed on low terraces with a range of slope less than 2 percent. The effective soil depth is very deep. The soil texture is loam over clayey loam grading to clay with few iron concretions in the subsoil. The surface is flooded by impounded rain water up to 50 cm deep for 3-4 months. The groundwater table is within 1 m depth for 5 months during the rainy season. This soil series is poorly drained with a moderate permeability and a slow surface runoff. This soil type found in the study area covers an area of 1,475.187 hectares, or 2.36 percent of the total area.

(5) Roi Et series, clayey variant (Re-c). This series is also an old riverine alluvial soil that was formed on low terraces with a range of slope less than 2 percent. The effective soil depth is very deep. The soil texture is sandy loam over clay in the subsoil. The surface is flooded by impounded rain water or stream water up to 50 cm deep for 2-3 months during the rainy

season. The groundwater table is found in 1.5 m depth in the dry season. This soil series is poorly drained with a moderate permeability and a slow surface runoff. This soil type found in the study area covers an area of 58.187 hectares, or 0.09 percent of the total area.

(6) Ubon series (Ub). This series is also an old riverine alluvial soil that was formed on low terraces with a range of slope less than 2 percent. The effective soil depth is deep. The soil texture is loamy sand over sandy loam or sand. The surface is flooded by impounded rain water up to 20 cm deep for 2-3 months during the rainy season. The groundwater level drops to 4 m depth in dry season. This soil series is somewhat excessively or well drained with a rapid permeability and a slow surface runoff. This soil type found in the study area covers an area of 200.062 hectares, or 0.32 percent of the total area.

(7) Roi Et series, high phase (Re-h). This series is an old riverine alluvial soil that was formed on low middle terraces with a range of slope 1-3 percent. The effective soil depth is very deep. The soil texture is sandy loam or sandy, clayey loam over clayey loam grading to clay. Rain water is impounded on the soil surface up to 25 cm for 3-4 months during the rainy season. The groundwater level falls below 3 m during dry season. This soil series is poorly drained with a rapid to moderate permeability and a slow surface runoff. This soil type found in the study area covers an area of 3,552.0 hectares, or 5.68 percent of the total area.

(8) Roi Et series, sandy variant (Re-s). This series is also an old riverine alluvial soil that was formed on low middle terraces with a gradient of about 1-2 percent. The effective soil depth is very deep. The soil texture is sandy loam or loamy sand over sandy loam at a depth greater than 80 cm. Rain water is impounded on the soil surface up to 25 cm for 3-4 months during the rainy season. The groundwater level falls below 3 m during dry season. This soil series is somewhat poorly drained with a moderate permeability and a slow surface runoff. This soil type found in the study area covers an area of 1,059.562 hectares, or 1.70 percent of the total area.

(9) Korat series (Kt). This series is an old riverine alluvial soil that was formed on middle terraces with a range of slope 2-6 percent. The effective soil depth is very deep. Soil texture is sandy loam over loam to sandy clay loam. The groundwater level is within 1.5 m below the surface during the peak of wet season. This soil series is moderately well drained with a moderate permeability and a rapid surface runoff. This soil type found in the study area covers an area of 7,122.937 hectares, or 11.40 percent of the total area.

(10) Korat series, sandy variant (Kt-s). This series is also an old riverine alluvial soil formed on middle terraces with a gradient of about 2-6 percent. The effective soil depth is very deep. The soil texture is sandy loam over sandy loam grading to sandy, clayey loam or sandy clay at some depth below 80 cm. The groundwater level is below 1.5 m all year round. This soil series is well drained with a rapid to moderate permeability and a rapid surface runoff. This soil type found in the study area covers an area of 962.687 hectares, or 1.54 percent of the total area.

(11) Korat series, shallow phase (Kt-sh). This series is the Korat series in the developing stage of a shallow phase. The effective soil depth is deep. This soil type found in the study area covers an area of 26.5 hectares, or 0.04 percent of the total area.

(12) Korat series, stony phase (Kt-st). This series is the Korat series in the developing stage of stony phase. This soil type found in the study area covers an area of 493.875 hectares, or 0.79 percent of the total area.

(13) Phon Phisai series (Pp). This series is an old riverine alluvial soil that was formed on middle terraces with a range of slope 2-6 percent. The effective soil depth is shallow or a layer of lateritic concretions. The soil texture is sandy loam or loam over sandy, clayey loam. The groundwater level is within 1 m below the surface during the wet season. This soil series is moderately well drained with a moderate to slow permeability and a rapid surface runoff. This soil type found in the study area covers an area of 9,220.750 hectares, or 14.75 percent of the total area.

(14) Korat/Phon Phisai association (Kt/Pp). This series is an association of the Korat and the Phon Phisai series. They are generally formed from the same parent materials and located continuously or alternately together in the same topographic area. According to the scale of the map used, these series cannot be easily differentiated. They are grouped together as an association. Soil characteristics depend on each associated soil series. This soil type found in the study area covers an area of 261.312 hectares, or 0.42 percent of the total area.

(15) Phen series (Pn). This series is an old riverine alluvial soil formed on middle terraces with a gradient of about 1-3 percent. The effective soil depth is shallow or a layer of lateritic concretions. The soil texture is sandy loam or loam over gravelly, sandy clay loam. Water at the surface is up to 30 cm for 2-3 months in the wet season. Groundwater table falls below 1 m during the dry season. This soil series is poorly drained with a moderate to slow permeability and a slow surface runoff. This soil type found in the study area covers an area of 879.125 hectares, or 1.41 percent of the total area.

(16) Nam Phong series (Ng). This series is also an old riverine alluvial soil that was formed on middle terraces with a range of slope between 3-10 percent. Effective soil depth is very deep. Soil texture is loamy sand over sand or loamy sand grading to sandy loam over sand. The groundwater table falls below 4 m during the dry season. This soil series is excessively drained with high permeability and rapid surface runoff. This soil type found in the study area covers an area of 13,880.0 hectares, or 22.21 percent of the total area.

(17) Nam Phong/Phon Phisai association (Ng/Pp). This series is the association of the Nam Phong and the Phon Phisai series. They are usually formed from the same parent materials and located continuously or alternately together in the same topographic area. According to the scale of the map used, these series cannot be easily differentiated. They are grouped together as an association. Soil characteristics depend on each associated soil series. It covers an area of 56.562 hectares, or 0.09 percent of the total area.

(18) Satuk series (Suk). This series is an old riverine alluvial soil formed on middle to high terraces with a slope ranging between 3-8 percent. Effective soil depth is very deep. Soil texture is loamy sand or sandy loam over sandy clay loam grading to sandy clay. The groundwater table lies below 1.5 m for 12 months. This soil series is well drained with a moderate permeability and a rapid surface runoff. This soil type found in the study area covers an area of 294.437 hectares, or 0.47 percent of the total area.

(19) Slope Complex (SC). Slope complex found in this study area covers an area of 3,147.312 hectares, or 5.04 percent of the total area.

(20) Rock-outcrop. Most of them are sandstone rocks. This type covers an area of 3,419.375 hectares, or 5.47 percent of the total area.

Additionally, some parts of this study area are water reservoirs that cover an area of 2,106.062 hectares, or 3.37 percent of the total area.

Figure 8 shows the distribution of soil units found in the study area.

Study area IV (Coastal zone). Based on the classification of the Department of Land Development, Bangkok, in 1981, soil types found in this study area can be identified as follows:

(1) Rayong series (Ry). This series is found in plain or undulating terrain with a gradient of about 2-4 percent. The effective soil depth is very deep. The soil texture is sandy throughout. This soil series is well drained with a high permeability. This soil type found in the study area covers an area of 811.812 hectares, or 1.30 percent of the total area.

(2) Ban Thon series (Bh). This series is found in undulating terrain with a gradient of about 2-4 percent. The effective soil depth is very deep. The soil texture is sandy or sandy loam. This soil series is moderately well drained with a moderate permeability. This soil type found in the study area covers an area of 240.437 hectares, or 0.38 percent of the total area.

(3) Rayong/Ban Thon association (Ry/Bh). This is an association of the Rayong and the Ban Thon series. They are generally formed from the same parent materials and located continuously or alternately together in the same topographic area. According to the scale of the map used, these series cannot be easily differentiated. They are grouped together as an association. The soil characteristics depend on each associated soil series. This soil type found in the study area covers an area of 670.812 hectares, or 1.07 percent of the total area.

(4) Tha Chin series (Tc). This series is formed on tidal flats, former tidal flats, and depressions between beach ridges. The topography of the areas where this soil series occurs is flat with a gradient of slope less than 1 percent. The parent materials of soil are marine deposits. The effective soil depth is very deep. The soil texture is sandy clay or clay in the upper layer and clay in the lower layer. The upper layer is brown with mottled deep gray. The lower layer is gray and deep green. This soil series is very poorly drained with a high permeability. This soil

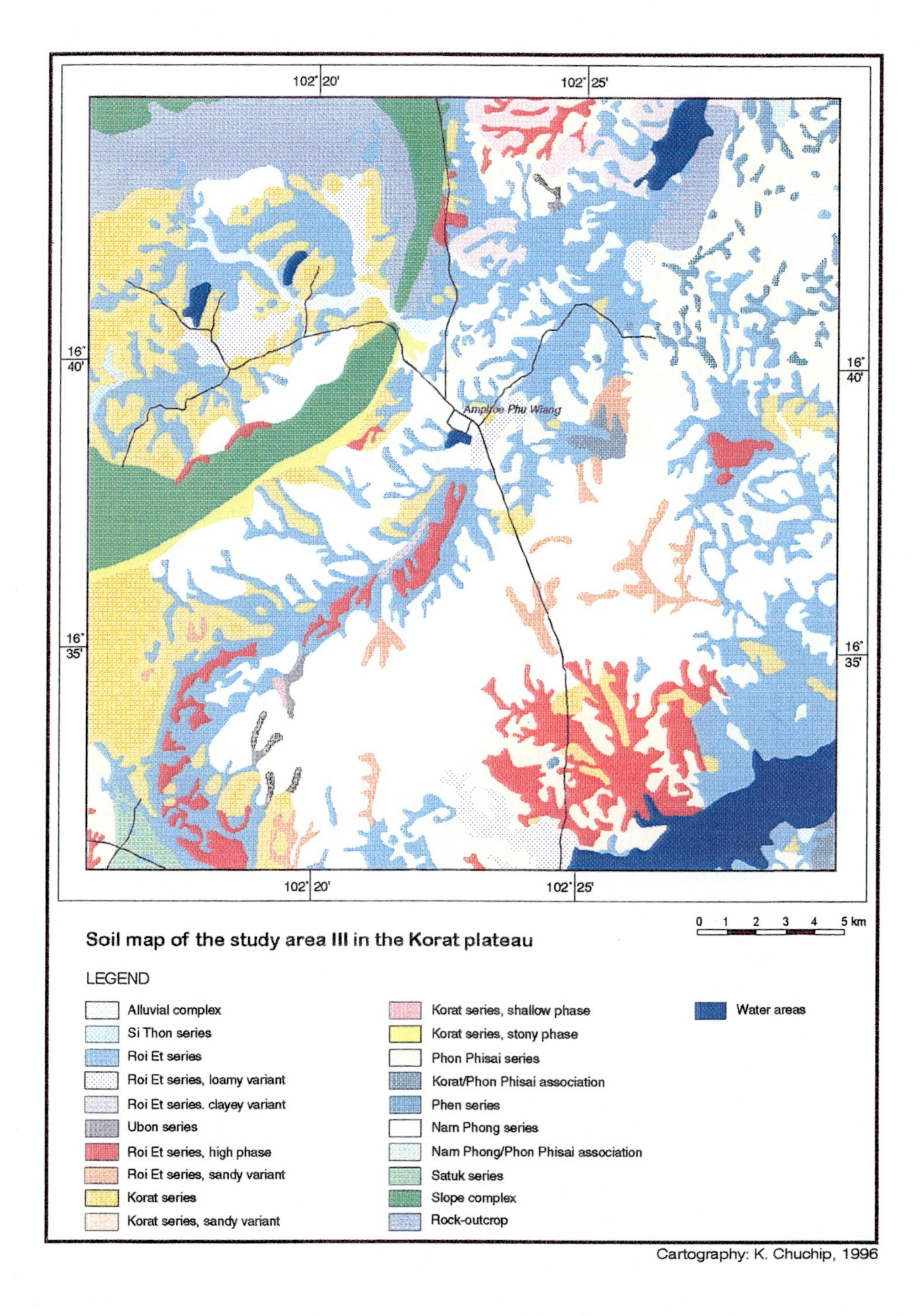

Fig. 8: **Distribution of soil units of the study area III in the Korat plateau (Source: Prepared by the author, based on the reconnaissance soil map scale to 1:100,000 of the Department of Land Development, 1973).**

type found in the study area covers an area of 14,674.687 hectares, or 23.48 percent of the total area.

(5) Bang Pakong seires (Bpg). This series is also formed on tidal flats, former tidal flats, and depressions between beach ridges. The topography of the areas where this soil series occurs is flat with a gradient of slope less than 1 percent. The parent materials of soil are marine deposits. This soil type is commonly acidic. The effective soil depth is very deep. The soil texture is clay both in the upper and the lower layer. The upper layer is brown to gray. The lower layer is gray or reddish brown. This soil series is very poorly drained with a high permeability. This soil type found in the study area covers an area of 2,465.562 hectares, or 3.94 percent of the total area.

(6) Samut Prakan series (Sm). This series is also formed on tidal flats, former tidal flats, and depressions between beach ridges. The topography of the areas where this soil series occurs is flat with a gradient of slope less than 1 percent. The parent materials of the soil are marine deposits. The effective soil depth is very deep. The soil texture is clay both in the upper and the lower layer. The upper layer is brown with mottled gray. The lower layer is gray or deep green with mottled deep brown. This soil series is poorly drained with a slow permeability. This soil type found in the study area covers an area of 1,025.875 hectares, or 1.64 percent of the total area.

(7) Cha Am series (Ca). This series is formed on former tidal flats, and swamps. The parent materials of this soil type are derived from brackish deposits. The topography of the areas where this soil series occurs is flat or relatively flat with a gradient of slope less than 1 percent. Effective soil depth is very deep. The soil texture is clay both in the upper and the lower layer. The upper layer is grayish brown with mottled gray. The lower layer is yellowish gray or yellowish red. This soil series is poorly drained. This soil type found in the study area covers an area of 5,943.750 hectares, or 9.51 percent of the total area.

(8) Ongkharak series (Ok). This series is also formed on former tidal flats and swamps. The parent materials of this soil type are derived from brackish deposits. The topography of the areas where this soil series occurs is flat with a gradient of slope less than 1 percent. The effective soil depth is very deep. The soil textures are clay or loamy clay in the upper layer and clay or loamy clay in the lower layer. This soil series is poorly drained with a low permeability. This soil type found in the study area covers an area of 826.375 hectares, or 1.32 percent of the total area.

(9) Alluvial Complex (AC). The parent materials of this soil type are recent and sub-recent riverine alluviums. It is normally formed on flood plains and valley flat. This soil type found in the study area covers an area of 379.937 hectares, or 0.61 percent of the total area.

(10) Ban Khai series (Bi). This series is normally formed on levees and flood plains. The parent materials of this soil type are derived from recent riverine alluvium. The topography of the areas where this soil series occurs is relatively flat with a gradient of slope less than 1 percent. The effective soil depth is very deep. The soil textures are clay or loamy sand to clayey loam in

the upper and the lower layer. The upper layer is brown to greyish brown. The lower layer is greyish brown with mottled yellowish brown. This soil type found in the study area covers an area of 60.062 hectares, or 0.10 percent of the total area.

(11) Bang Len series (Bl). This series is formed on low terraces and fans. The parent materials of this soil are derived from old alluviums. The topography of the areas where this soil series occurs is flat with a gradient of slope less than 1 percent. The effective soil depth is deep. The soil texture is clay both in the upper layer and in the lower layer. The upper layer has a pale color and mottled yellowish red . The lower layer is light gray. This soil series is relatively poor drained with a low permeability. This soil type found in the study area covers an area of 39.750 hectares, or 0.06 percent of the total area.

(12) Klaeng series (Kl). This series is also formed on low terraces and fans. The parent materials of this soil series are derived from old alluviums. The topography of the areas where this soil series occurs is nearly flat with a gradient of slope less than 1 percent. The effective soil depth is very deep. The soil textures are sandy clay loam or loamy clay in the upper layer and clay or sandy clay in the lower layer. This soil series is poorly drained and of low permeability. This soil type found in the study area covers an area of 634.375 hectares, or 1.01 percent of the total area.

(13) Stul (Stu). This series is formed on low terraces and fans. The parent materials of this soil are derived from old alluviums. The topography of the areas where this soil series occurs is flat with a gradient of slope less than 2 percent. Effective soil depth is very deep. Soil texture are sandy loam in upper layer and sandy clay in lower layer. This soil series is poorly drained and of low permeability. This soil type found in the study area covers an area of 5.375 hectares, or 0.01 percent of the total area.

(14) Khlaeng/Soldic soil association (Kl/Sol). This series is formed on low terraces and fans. This series is composed of Khlaeng series (Kl) and Soldic soil. The soil parent materials are derived from old alluviums and located continuously or alternately together in the same topographic area. According to the scale of the map used, these series cannot be easily differentiated. They are grouped together as an association. Soil characteristics depend on associated soil series. This soil type found in the study area covers an area of 2.437 hectares, or 0.004 percent of the total area.

(15) Chon Buri series (Cb). This series is also formed on low terraces and fans. Parent materials of this soil are derived from old alluviums. The topography of the areas where this soil series occurs is relatively flat with a gradient of slope less than 2 percent. Effective soil depth is very deep. Soil textures are sandy loam or sandy clay in the upper layer and sandy clay loam to sandy clay in the lower layer. This soil series is poorly drained with a moderately high permeability. This soil type found in the study area covers an area of 1,532.687 hectares, or 2.45 percent of the total area.

(16) Nam Krachai series (Ni). This series is also formed on low terraces and fans. Parent materials of this soil are derived from old alluviums. The topography of the areas where this

soil series occurs is flat with a gradient of slope less than 2 percent. Effective soil depth is very deep. Soil textures are loamy sand or sandy loam in the upper layer and sandy loam or sandy clay loam in the lower layer. This soil series is poorly drained and of moderate permeability. This soil type found in the study area covers an area of 1,120.812 hectares, or 1.79 percent of the total area.

(17) Nam Krachai, gravelly subsoil variant (Ni-gr). This series is also formed on low terraces and fans. The parent materials of this soil are derived from old alluviums. The topography of the areas where this soil series occurs is relatively flat with a gradient of about 2-4 percent. The effective soil depth is moderately deep. Soil textures are loamy sand or sandy loam in the upper layer and loamy clay mixed with gravelly sand in the lower layer. This soil series is poorly drained with a moderate permeability. This soil type found in the study area covers an area of 304.250 hectares, or 0.49 percent of the total area.

(18) Visai series (Vi). This series is also formed on low terraces and fans. The parent materials of this soil type are derived from old alluviums. The topography of the areas where this soil series occurs is nearly flat with a gradient of slope less than 2 percent. Effective soil depth is very deep. Soil textures are sandy clay loam in the upper layer and sandy clay or loamy clay in the lower layer. This soil series is poorly drained with a moderate permeability. This soil type found in the study area covers an area of 401.125 hectares, or 0.64 percent of the total area.

(19) Sungai Padi series (Pi). This series is also formed on low terraces and fans. The parent materials of this soil type are derived from old alluviums. The topography of the areas where this soil series occurs is nearly flat to undulating with a gradient of slope less than 2 percent. The effective soil depth is very deep. Soil textures are sandy loam or sandy clay loam in the upper layer and loamy clay mixed with gravelly sand in the lower layer. This soil series is poorly drained with a moderate permeability. This soil type found in the study area covers an area of 214.562 hectares, or 0.34 percent of the total area.

(20) Na Thawi series (Nat). This series is formed on middle terraces. The parent materials of this soil are derived from old alluviums. The topography of the areas where this soil series occurs is undulating with a gradient of slope ranging between 3-8 percent. The effective soil depth is very deep. Soil textures are loamy sand or sandy loam in the upper layer and sandy clay loam in the lower layer. This soil series is relatively well drained with a high permeability. This soil type found in the study area covers an area of 22.375 hectares, or 0.04 percent of the total area.

(21) Chumphon series (Cp). This series is also formed on middle terraces. The parent materials of this soil type are derived from old alluviums. The topography of the areas where this soil series occurs is undulating to steeply undulating with a gradient of slope ranging between 3-8 percent. The effective soil depth is deep. Soil textures are sandy loam in the upper layer and gravelly clay loam or gravelly clay in the lower layer. This soil series is relatively well drained with a moderate permeability. This soil type found in the study area covers an area of 158.062 hectares, or 0.25 percent of the total area.

(22) Hat Yai series (Hy). This series is also formed on middle terraces. The parent materials of this soil type are derived from old alluviums. The topography of the areas where this soil series occurs is undulating. The effective soil depth is deep. Soil textures are sandy loam or sandy clay loam in the upper layer and gravelly clay loam or gravelly clay in the lower layer. This soil series is well drained with a moderate permeability . This soil type found in the study area covers an area of 19.437 hectares, or 0.03 percent of the total area.

(23) Huai Pong series (Hp). This series is formed on coalescing fans. The parent materials of this soil type are derived from old alluviums and colluviums. The topography of the areas where this soil series occurs is nearly flat to undulating with a gradient of slope ranging between 2-6 percent. The effective soil depth is very deep. Soil textures are sandy clay loam or sandy loam in the upper layer and sandy clay loam or sandy clay in the lower layer. This soil series is well drained with a moderate permeability. This soil type found in the study area covers an area of 1,878.187 hectares, or 3.01 percent of the total area.

(24) Huai Pong, gravelly subsoil variant (Hp-gr). This series is also formed on coalescing fans. The parent materials of this soil type are derived from old alluviums and colluvium. The topography of the areas where this soil series occurs is nearly flat to undulating with a gradient of slope ranging between 2-6 percent. The effective soil depth is moderately deep. Soil textures are sandy clay loam or sandy loam in the upper layer and gravelly sandy clay loam in the lower layer. This soil series is well drained and of moderate permeability. This soil type found in the study area covers an area of 50.312 hectares, or 0.08 percent of the total area.

(25) Khlong Chak series (Kc). This series is an erosional surface. The parent materials of this soil type are derived from residuum and colluvium. The topography of the areas where this soil series occurs is undulating to hilly with a gradient of slope ranging between 4-12 percent. The effective soil depth is shallow to moderately deep. Soil textures are gravelly clay loam or loam in the upper layer and gravelly clay loam or gravelly clay in the lower layer. This soil series is well drained with a high permeability. This soil type found in the study area covers an area of 2,343.750 hectares, or 3.75 percent of the total area.

(26) Trang series (Tng). This series is formed on hills and foot-slopes. The parent materials of this soil type are derived from residuum and local colluvium. The topography of the areas where this soil series occurs is undulating to steeply undulating, or foot-slopes with a gradient of slope ranging between 4-16 percent. The effective soil depth is deep. Soil textures are clayey loam or clay in the upper layer and clay in the lower layer. This soil series is well drained with a moderate permeability. This soil type found in the study area covers an area of 108.562 hectares, or 0.17 percent of the total area.

(27) Trad series (Td). This series is also formed on hills and footslopes. The parent materials of this soil type are derived from residuum and local colluvium. The topography of the areas where this soil series occurs is undulating to steeply undulating, or foot-slopes with a gradient of slope ranging between from 2-6 percent. The effective soil depth is deep. Soil textures are loam or clay loam in the upper layer and gravelly clay loam or gravelly clay in the lower layer.

This soil series is well drained with a high permeability. This soil type found in the study area covers an area of 1,023.125 hectares, or 1.64 percent of the total area.

(28) Trad/Khlong Chak association (Td/Kc). This series is also formed on hills and footslopes. This series is composed of the Trad (Td) and the Khlong Chak (Kc) series. They are generally formed from the same parent materials and located continuously or alternately together in the same topographic area. According to the scale of the map used, these series cannot be easily differentiated. They are grouped together as an association. Soil characteristics depend on each associated soil series. This soil type found in the study area covers an area of 1,265.750 hectares, or 2.03 percent of the total area.

(29) Phuket, yellow variant (Pk-y). This series is also an erosional surface. The parent materials of this soil type are derived from residuums and colluviums. The topography of the areas where this soil series occurs is undulating to steeply undulating with a gradient of slope ranging between 4-20 percent. The effective soil depth is deep. Soil textures are sandy clay loam in the upper layer and sandy clay or clay with a few gravel in the lower layer. This soil series is well drained with a moderate permeability. This soil type found in the study area covers an area of 332.250 hectares, or 0.53 percent of the total area.

(30) Ranong series (Rg). This series is formed on hills and foot-slopes. The parent materials of this soil type are derived from residuum and colluvium. The topography of the areas where this soil series occurs are foot-slopes or isolated hills with a gradient of slope ranging between 8-30 percent. The effective soil depth is shallow to deep. Soil textures are sandy loam or loam in the upper layer and gravelly loam or gravelly clay loam in the lower layer. This soil series is well drained with a high permeability. This soil type found in the study area covers an area of 707.0 hectares, or 1.13 percent of the total area.

(31) Ranong/Phato association (Rg/Pto). This series is also formed on hills and foot-slopes. The parent materials of this soil type are derived from residuum and colluvium. The topography of the areas where this soil series occurs is steeply undulating, or foot-slopes with a gradient of slope ranging between 8-30 percent. The effective soil depth is shallow to deep. Soil textures are sandy loam or loam in the upper layer and gravelly loam or gravelly clay loam in the lower layer. This soil series is well drained with a high permeability. This soil type found in the study area covers an area of 101.312 hectares, or 0.16 percent of the total area.

(32) Slope Complex (SC). This is a definition of lands belonging to mountainous areas with a gradient of more than 35 percent. Most of the areas covered with forests. The slope complex found in this study area covers an area of 1,015.375 hectares, or 1.62 percent of the total area.

(33) Unnamed soil (U). Soil in some areas, especially on islands, have not been identified yet. This type found in the study area is confined to two small islands that cover an area of 78.562 hectares, or 0.13 percent of the total area. Most of these areas are covered with moist evergreen forests. Some parts are beaches.

(34) Mud. The coastlines of this study area are generally covered with mud-flats. This zone is very important for the coastal ecology. Mud-flats of this area are seed beds for mangrove forests. This type covers an area of 4,721.5 hectares, or 7.55 percent of the total area.

(35) Water. Around half of the study area is covered with water surfaces from rivers and a part of the Gulf of Thailand. Water covers an area of 17,319.750 hectares, or 27.71 percent of the total area.

Figure 9 shows the distribution of soil units found in the study area.

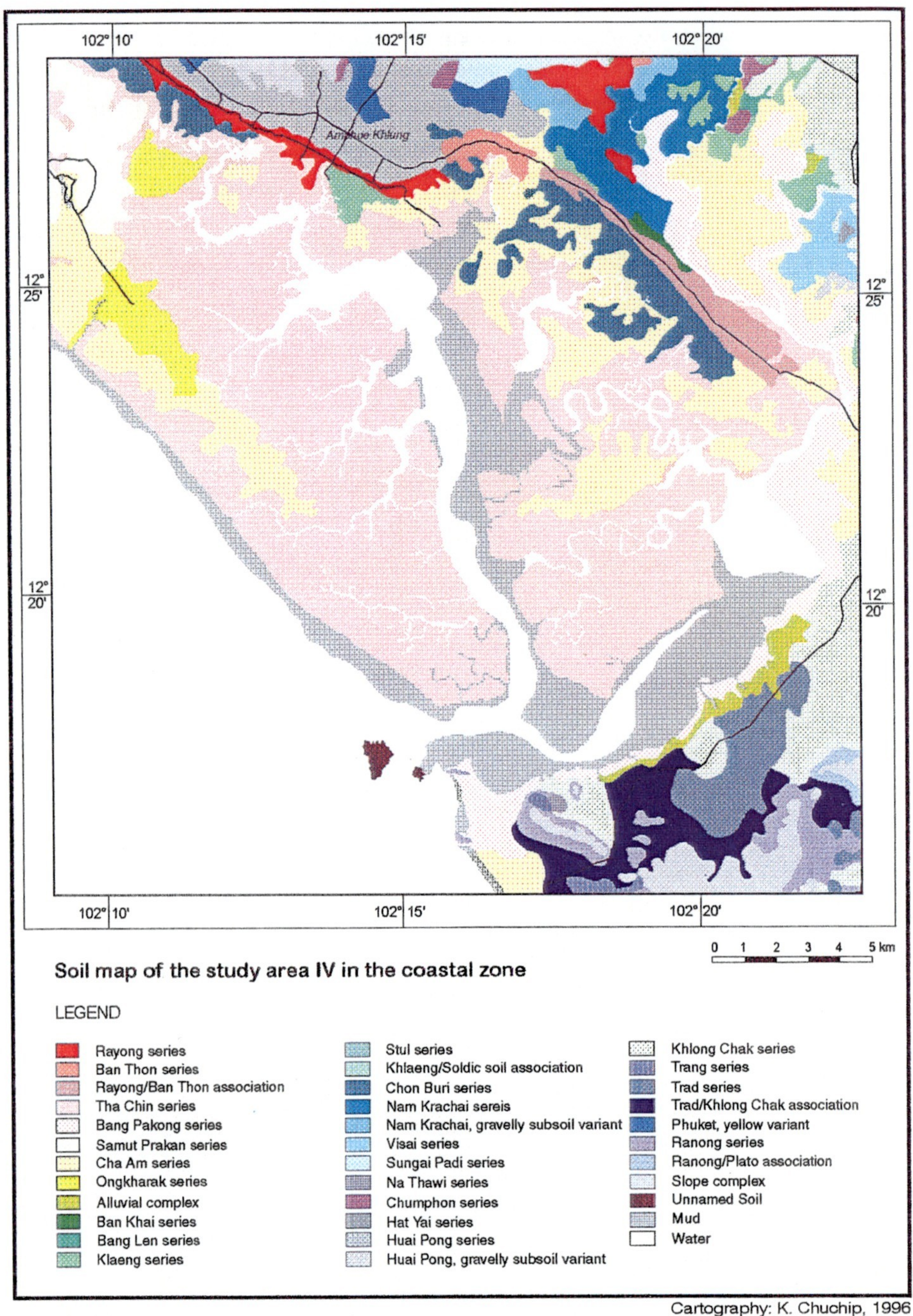

Fig. 9: **Distribution of soil units of the study area IV in the coastal zone (Source: Prepared by the author, based on the reconnaissance soil map scale to 1:100,000 of the Department of Land Development, 1973).**

4.5 The general climate of the study areas

Thailand lies entirely within the tropical zone, and is under the influence of seasonal monsoon winds: Seasonal influences of the Pacific Ocean trade winds and the Asiatic monsoons, the Northeast monsoon and the Southwest monsoon, result in a climate showing two distinct seasons. These seasons are the dry season, mainly between mid-October and March, and rainy season, mainly between mid-May and mid-September. In addition, short transition periods separate these monsoonal seasons into four seasons. Climatic variations in Thailand depend mainly upon the variation of rainfall. However, locally the higher relief influences the climatic conditions of each area.

Northeast Monsoon: The Northeast monsoon or trade winds normally begins around the middle of October. The cooler and drier air from the Northeast dominates the area beginning from the Northeast and the North of the country. Cloudiness is normally considerably less than during the Southwest monsoon period. Clear days are much more frequent, but cumulus clouds build up in the afternoon. Early morning fog is common in deeper river valleys, but usually dissipates by noon. Excellent visibility is rare because of persistent haze. Nighttime temperatures are noticeably lower, but in the daytime they are still relatively high.

Spring Transition Period. The northeasterly flow of air dissipates by March and does not longer dominate the area. During this transition period, just before the Southwest monsoon starts, temperatures reach their annual maximum. Afternoon temperatures are higher. The humidity is still low as also the cloud coverage. The sky remains relatively clear.

Southwest Monsoon: The Southwest monsoon or rainy season usually begins in the middle of May and is well established by June. Low clouds, heavy rain-showers, and thunderstorms are prevalent. The onset of this southwesterly flow is marked by heavy cumulonimbus clouds, squalls, and severe thunderstorms. A nearly regular pattern is established by June, with almost daily local showers, occasionally torrential, occurring during the afternoon and early evening. These showers are caused mainly by convection. During squalls and heavy showers, the bases of clouds are quite low and the visibility is greatly reduced for short periods. During August and September developing tropical storms enter the area infrequently with longer lasting and heavy rainfall. Periods of fair weather with scattered clouds occur occasionally, but clear days are rare. Surface visibility is fairly good, except during periods of precipitation. The temperature decreases, especially in the higher elevations. The high temperatures are accompanied by high humidity.

Autumn Transition Period: During the early part of the autumn transition period in the middle of September, the weather is quite similar to that of the Southwest monsoon. There is some decrease in cloudiness, rainfall, and local storms. Temperatures remain about the same as during the Southwest monsoon period, although in many locations daytime maximums are slightly higher. Temperatures remain about the same as during the Southwest monsoon period. By mid-October, the drier and cooler northeasterly flow dominates the area.

Geographisches Institut
der Universität Kiel

4.5.1 RAINFALL

Study area I (Mountainous area). The climatological data available in the study area, recorded from 1981-1990, are shown in Tab. 8. The data were recorded at the station 328003, Lampang province. The station is situated at 18° 42' latitude and 100° 00' longitude. It is 241 m above MSL high. Over the period of observation, the average annual rainfall is 1,039.0 mm with 65.1 annual average rainy days. Although the total amount of rainfall is not quite high, the area has relatively regular Rainy days from April to November. Figure 10 shows the monthly rainy days over that period. From data analysis, it can be concluded that the rainy season occurs from April to November and has a bimodal distribution with the first peak in April to June and the second in July to November, as shown in Fig. 11. The early rainy season (April-June) is mainly derived from the Southwest Monsoon. The total rainfall in this period is usually less than in the second peak (July-November). The second rainy period is caused by the South West monsoon and cyclonic rains from the South China Sea. Rains during this second period are more frequent and heavier than in the first peak.

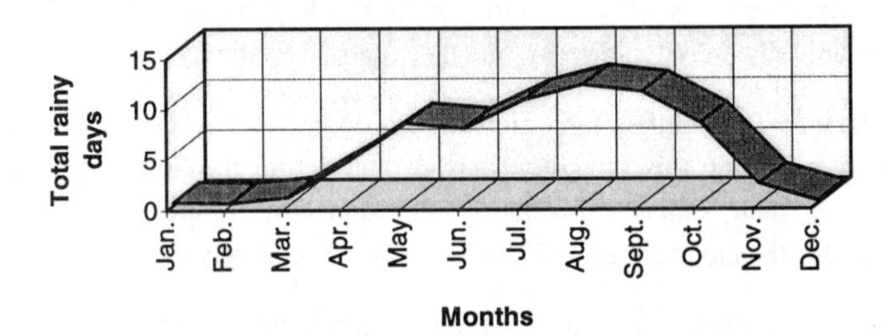

Fig. 10: **Monthly total rainy days during the period between 1981 and 1990 of the study area I (Source: Prepared by the author using the data of the Meteorological Department, Bangkok).**

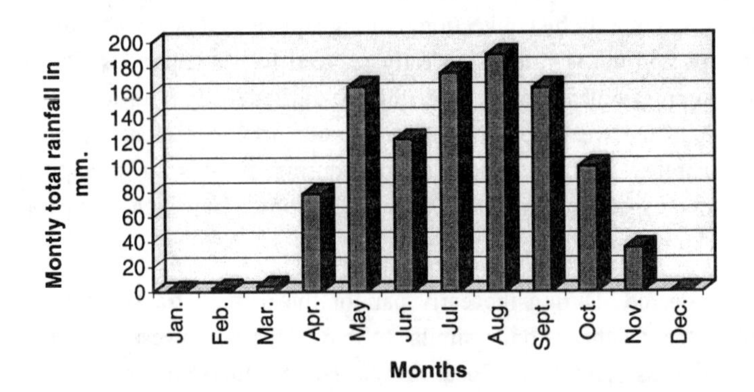

Fig. 11: **Monthly total rainfall (in mm) during the period between 1981 and 1990 of the study area I (Source: Prepared by the author using the data of the Meteorological Department, Bangkok).**

Tab. 8: **Monthly rainfall and rainy days over the study area I (in the mountainous area) between 1981 and 1990, measured at the Station 328003 (Location: 18° 42' Lat., 100°00' Long., 241.0 m above MSL).**

Rainfall: in mm

Year	Monthly total	Jan.	Feb.	Mar.	Apr.	May	Jun.	Jul.	Aug.	Sep.	Oct.	Nov.	Dec.	**Annual**
1981	Rainfall	1.8	0.0	0.4	65.6	256.3	73.5	394.6	220.3	151.2	161.3	51.9	1.2	**1,378.1**
	Rainy days	1	0	1	6	12	12	18	16	10	16	5	1	**98**
1982	Rainfall	3.4	0.0	0.0	111.2	127.7	76.0	158.8	133.1	150.0	75.5	9.4	0.0	**845.1**
	Rainy days	1	0	0	10	10	11	13	18	16	9	2	0	**90**
1983	Rainfall	1.9	0.0	0.9	24.5	96.0	113.1	202.2	127.0	150.5	135.3	56.8	5.2	**913.4**
	Rainy days	1	0	1	1	9	10	10	16	14	9	3	2	**76**
1984	Rainfall	0.0	9.6	1.0	124.5	158.7	175.4	94.8	198.4	214.9	96.8	0.0	0.0	**1,074.1**
	Rainy days	0	1	1	5	8	7	10	14	10	6	0	0	**62**
1985	Rainfall	0.0	0.0	0.0	148.7	160.4	166.2	166.9	126.8	252.5	92.5	74.9	0.0	**1,188.9**
	Rainy days	0	0	0	6	6	7	4	9	14	5	4	0	**55**
1986	Rainfall	0.0	0.0	0.0	60.6	189.2	91.3	209.7	-	116.9	154.6	0.0	0.0	**-**
	Rainy days	0	0	0	3	7	5	11	-	7	10	0	0	**-**
1987	Rainfall	0.0	0.0	0.0	45.3	138.2	168.7	90.3	317.6	201.7	43.8	114.5	0	**1,120.1**
	Rainy days	0	0	0	3	5	7	8	12	11	2	1	0	**49**
1988	Rainfall	0.0	0.0	0.0	71.8	98.9	159.5	90.7	242.7	90.5	96.4	30.2	0.0	**880.7**
	Rainy days	0	0	0	3	4	8	5	7	7	8	3	0	**45**
1989	Rainfall	0.0	0.0	0.0	74.5	277.6	103.1	169.9	182.9	-	71.6	0.0	0.0	**-**
	Rainy days	0	0	0	1	8	3	7	6	-	10	0	0	**-**
1990	Rainfall	0.0	15.1	43.5	58.5	142.0	71.4	176.5	160.8	145.4	78.2	16.5	0.0	**907.9**
	Rainy days	0	1	5	4	12	6	18	9	12	6	2	0	**75**
Mean	**Rainfall**	**0.7**	**2.5**	**4.6**	**78.5**	**164.5**	**122.5**	**175.4**	**190.0**	**163.7**	**100.6**	**35.4**	**0.6**	**1,039.0**
Mean	**Rainy days**	**0**	**1**	**5**	**4**	**12**	**6**	**18**	**9**	**12**	**6**	**2**	**0**	**75**

Source: METEOROLOGICAL DEPARTMENT, 1992

Remarks: 'T' means trace rainfall with having an amount of less than 0.1 mm.

Study area II (Central plain). The climatological data available in the study area recorded from 1981-1990 are shown in Tab. 9. The data were recorded at station 412001, Angthong province. The station is situated at 14° 35' latitude and 100° 27' longitude. It is about 8 m above MSL high. Over this period, the average annual rainfall is 1,208.0 mm with 85.9 annual average rainy days. Noticeably, the area has rainy days all year. September has most frequent rainy days in this area. Figure 12 shows the monthly rainy days over that period. From data analysis, it can be concluded that the rainy season occurs from May to October. It has a bimodal distribution with the first highest peak in May and the second highest peak in September, as shows in Fig. 13. The early rainy season (May-June) is mainly derived from the Southwest Monsoon. The total rainfall in this period is usually less than in the second peak (July-October). The second period of rainy days is derived from the South West monsoon and cyclonic rain from the South China sea. Rains during this second period are more frequent and heavier than during the first peak.

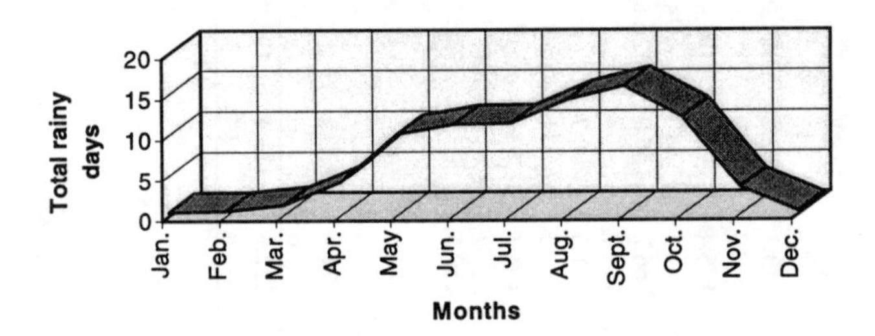

Fig. 12: **Monthly total rainy days during the period between 1981 and 1990 of the study area II (Source: Prepared by the author using the data of the Meteorological Department, Bangkok).**

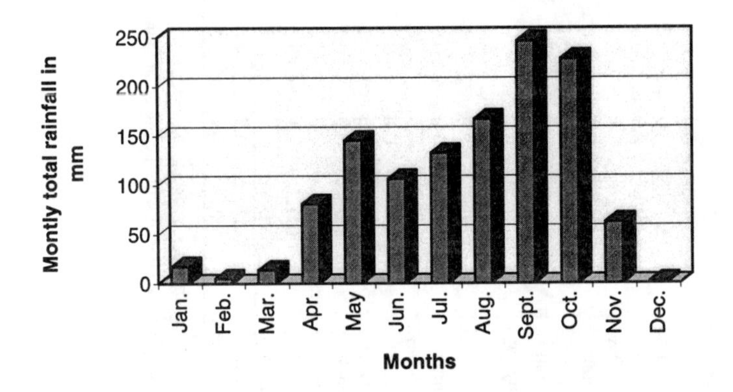

Fig. 13: **Monthly total rainfall (in mm) during the period between 1981 and 1990 of the study area II (Source: Prepared by the author using the data of the Meteorological Department, Bangkok).**

Tab. 9: **Monthly rainfall and rainy days in the study area II (in the Central Plain) between 1981 and 1990, measured at the Station 412001 (Location: 14° 35' Lat., 100°27' Long., 8.0 m above MSL).**

Rainfall: in mm

Year	Monthly total	Jan.	Feb.	Mar.	Apr.	May	Jun.	Jul.	Aug.	Sep.	Oct.	Nov.	Dec.	**Annual**
1981	Rainfall	0.0	T	36.0	52.1	123.7	126.7	156.7	160.7	297.2	58.7	169.7	0.0	**1,181.5**
	Rainy days	0	0	2	4	13	13	15	14	18	6	7	0	**92**
1982	Rainfall	0.0	0.0	11.4	127.7	153.5	90.7	131.7	83.1	175.4	113.5	179.8	6.1	**1,072.9**
	Rainy days	0	0	2	4	10	12	8	17	14	12	6	1	**86**
1983	Rainfall	0.3	0.0	T	0.0	208.2	149.0	69.4	253.2	161.4	443.9	144.6	13.7	**1,443.7**
	Rainy days	1	0	0	0	8	13	10	21	15	20	6	2	**96**
1984	Rainfall	8.3	12.9	38.8	89.7	187.0	89.4	262.2	53.1	266.6	137.3	T	0.2	**1,145.5**
	Rainy days	1	2	3	5	8	11	14	12	17	11	0	1	**85**
1985	Rainfall	14.4	T	0.0	66.4	202.7	75.1	73.1	104.8	130.9	220.6	48.8	0.0	**936.8**
	Rainy days	1	0	0	4	9	10	12	12	16	14	5	0	**83**
1986	Rainfall	0.0	0.0	T	115.3	183.7	87.5	191.0	209.7	375.0	120.2	0.0	2.8	**1,285.2**
	Rainy days	0	0	0	9	10	11	14	14	14	9	0	1	**82**
1987	Rainfall	0.0	0.0	4.9	18.0	69.9	96.1	102.4	111.5	270.7	92.4	49.9	0.0	**815.8**
	Rainy days	0	0	3	5	5	15	8	136	18	11	6	0	**84**
1988	Rainfall	0.0	36.3	0.0	132.4	80.9	137.5	247.3	259.6	249.8	332.1	0.0	0.0	**1,475.9**
	Rainy days	0	3	0	7	16	13	16	18	12	14	0	0	**99**
1989	Rainfall	153.4	2.8	0.9	63.2	116.2	104.6	54.3	201.3	298.5	102.6	0.8	0.0	**1,098.6**
	Rainy days	1	1	1	2	10	6	8	13	19	11	1	0	**73**
1990	Rainfall	0.6	0.0	48.2	144.4	130.4	106.1	36.8	233.9	237.5	661.6	30.5	0.0	**1,630.0**
	Rainy days	1	0	2	2	12	9	9	9	17	14	4	0	**79**
Mean	**Rainfall**	**17.7**	**5.2**	**14.0**	**80.9**	**145.6**	**106.3**	**132.5**	**167.1**	**246.3**	**228.3**	**62.4**	**2.3**	**1,208.6**
Mean	**Rainy days**	**0.5**	**0.6**	**1.3**	**4.2**	**10.1**	**11.3**	**11.4**	**14.3**	**16.0**	**12.2**	**3.5**	**0.5**	**85.9**

Source: METEOROLOGICAL DEPARTMENT, 1992

Remarks: 'T' means trace rainfall with having an amount of less than 0.1 mm.

Study area III (Korat plateau). The climatological data available in the study area recorded from 1981-1990 are shown in Tab. 10. The data were recorded at station 381003. The station is located at 16° 29' latitude and 102° 07' longitude. It is about 170.0 m above MSL high. Over the period, the average annual rainfall is 1,062.6 mm with 63.4 annual average rainy days. Although the total amount of rainfall is not quite high, the area has relatively regular rainy days from May to October. Furthermore, the area has a very high rate of 'maximum rain a day' (20.76 mm) that occurred in October. This maximum rain is expected to have a high erosivity. Figure 14 shows the monthly rainy days over that period. From data analysis, it can be concluded that the rainy season occurs from May to October and has a bimodal distribution with the first peak in May to June and the second in July to October, as shows in Fig. 15. The early rainy season (May-June) is mainly derived from the Southwest monsoon. The total rainfall in this period is usually less than in the second peak (July-October). The second period of rainy days is derived from the South West monsoon and cyclonic rains from the South China Sea. Rains during this second period are more frequent and heavier than in the first peak

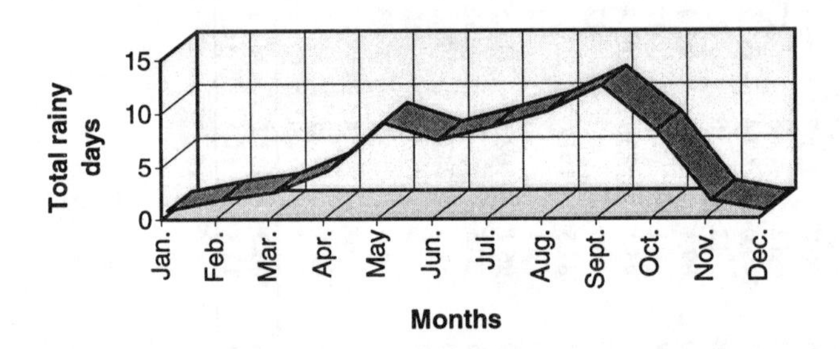

Fig. 14: **Monthly total rainy days during the period between 1981 and 1990 of the study area III (Source: Prepared by the author using the data of the Meteorological Department, Bangkok).**

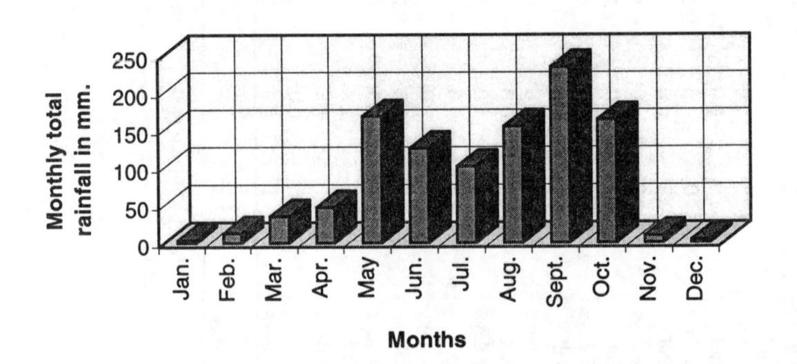

Fig. 15: **Monthly total rainfall (in mm) during the period between 1981 and 1990 of the study area III (Source: Prepared by the author using the data of the Meteorological Department, Bangkok).**

Tab. 10: Monthly rainfall and rainy days over the study area III (in the Korat Plateau) between 1981 and 1990, measured at the Station 381003 (Location: 16° 29' Lat., 102°07' Long., 170.0 m above MSL).

Rainfall: in mm

Year	Monthly total	Jan.	Feb.	Mar.	Apr.	May	Jun.	Jul.	Aug.	Sep.	Oct.	Nov.	Dec.	**Annual**
1981	Rainfall	0.0	T	10.5	14.0	45.0	78.5	85.2	50.1	133.4	156.9	20.0	0.0	**593.6**
	Rainy days	0	0	2	3	6	5	8	5	10	5	2	0	**46**
1982	Rainfall	0.0	0.0	24.7	51.3	182.6	109.2	122.1	99.3	297.1	177.7	0.0	28.7	**1,092.7**
	Rainy days	0	0	3	2	5	5	7	7	13	8	0	1	**51**
1983	Rainfall	11.4	0.0	T	0.0	161.1	349.4	145.4	282.2	151.8	142.1	7.9	T	**1,251.3**
	Rainy days	1	0	0	0	4	10	11	15	10	9	1	0	**61**
1984	Rainfall	13.0	21.4	44.2	48.6	143.0	172.6	81.0	118.7	232.1	138.2	T	0.0	**1,012.8**
	Rainy days	1	2	1	6	7	4	11	11	11	9	0	0	**63**
1985	Rainfall	2.6	10.0	T	35.4	217.1	68.7	146.1	96.5	207.1	207.0	2.1	0.0	**992.6**
	Rainy days	1	3	0	4	12	10	8	5	16	7	1	0	**67**
1986	Rainfall	0.0	0.0	10.8	23.8	184.4	95.7	82.3	113.7	176.6	56.9	4.2	9.1	**757.5**
	Rainy days	0	0	1	4	13	7	6	8	9	5	1	2	**56**
1987	Rainfall	0.0	28.4	108.5	47.2	122.9	59.5	44.6	294.5	253.6	37.4	38.8	0.0	**1,035.4**
	Rainy days	0	2	6	7	7	9	5	13	16	5	6	0	**76**
1988	Rainfall	0.0	24.3	1.3	127.2	193.6	152.7	150.2	170.2	166.4	396.0	0.0	0.0	**1,381.9**
	Rainy days	0	3	1	10	13	9	14	11	10	16	0	0	**87**
1989	Rainfall	10.2	0.0	55.1	40.6	187.8	94.1	68.0	163.4	250.5	189.9	0.0	0.0	**1,059.6**
	Rainy days	1	0	4	4	7	6	6	11	12	8	0	0	**59**
1990	Rainfall	0.0	45.6	98.8	86.2	247.5	78.3	92.0	162.1	474.0	137.4	24.8	0.0	**1,446.7**
	Rainy days	0	4	3	2	13	5	7	11	14	7	2	0	**68**
Mean	**Rainfall**	**3.7**	**13.0**	**35.4**	**47.4**	**168.5**	**125.9**	**101.7**	**155.1**	**234.3**	**164.0**	**9.8**	**3.8**	**1,062.6**
Mean	**Rainy days**	**0.4**	**1.4**	**2.1**	**4.2**	**8.7**	**7.0**	**8.3**	**9.7**	**12.1**	**7.9**	**1.3**	**0.3**	**63.4**

Source: METEOROLOGICAL DEPARTMENT, 1992

Remarks: 'T' means trace rainfall with having an amount of less than 0.1 mm.

Study area IV (Coastal zone). The climatological data available in the study area recorded from 1981-1990 are shown in Tab. 11. The data were recorded at the station 480003, Chantaburi province. The station is located at 12° 22' latitude and 102° 21' longitude. It is about 3.0 m above MSL high. Over the period, the average annual rainfall is 3,024.4 mm with 124.8 annual average rainy days. Noticeably, the area has rainy days all year and has also the great amount of annual rainfall. September has most frequent rainy days in this area. Figure 16 shows the monthly rainy days over that period. From data analysis, it can be concluded that the rainy season occurs from May to October. It also has a bimodal distribution with the first peak in May to June and the second peak in July to October, as shows in Fig. 17. However, it has usually been raining from May to October in this area.

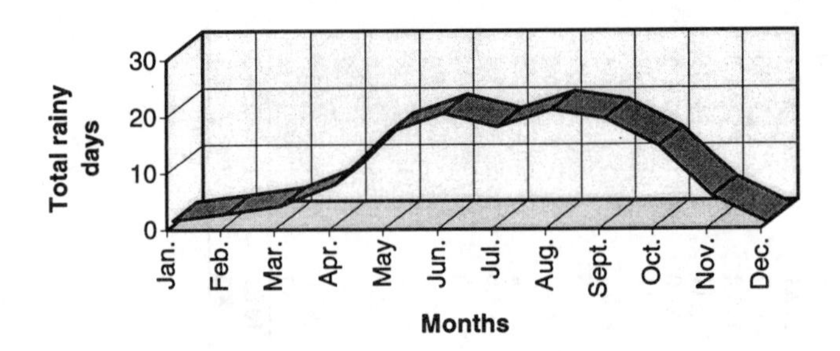

Fig. 16: **Monthly total rainy days during the period between 1981 and 1990 of the study area IV (Source: Prepared by the author using the data of the Meteorological Department, Bangkok).**

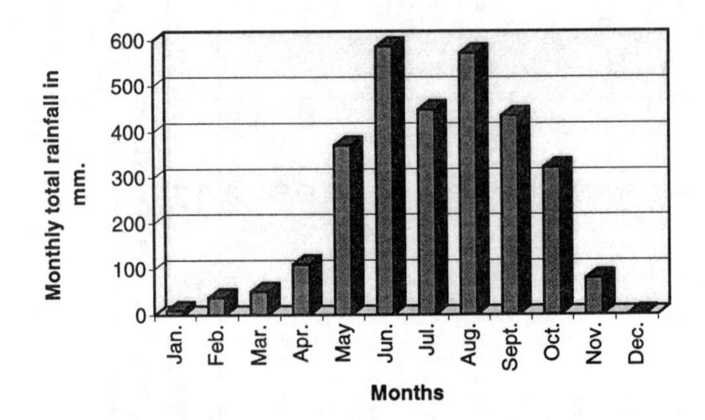

Fig. 17: **Monthly total rainfall (in mm) during the period between 1981 and 1990 of the study area IV (Source: Prepared by the author using the data of the Meteorological Department, Bangkok).**

Tab. 11: **Monthly rainfall and rainy days over the study area IV (in the coastal zone) between 1981 and 1990, measured at the Station 480003 (Location: 12° 22' Lat., 102°21' Long., 3.0 m above MSL).**

Rainfall: in mm

Year	Monthly total	Jan.	Feb.	Mar.	Apr.	May	Jun.	Jul.	Aug.	Sep.	Oct.	Nov.	Dec.	**Annual**
1981	Rainfall	0.0	222.3	12.6	92.0	277.2	527.9	631.3	559.1	650.8	289.3	184.6	0.0	**3,447.1**
	Rainy days	0	4	2	7	19	18	27	22	21	14	7	0	**141**
1982	Rainfall	0.0	9.8	50.6	227.7	177.1	704.3	683.5	690.2	536.2	193.0	124.0	0.0	**3,396.4**
	Rainy days	0	1	5	10	11	24	21	20	20	11	7	0	**130**
1983	Rainfall	0.0	26.7	0.0	5.0	475.9	494.6	577.3	1079.5	555.0	749.5	134.0	0.0	**4,097.5**
	Rainy days	0	1	0	1	14	19	19	26	15	17	5	0	**117**
1984	Rainfall	33.5	7.2	85.9	160.6	493.2	794.9	369.8	364.5	432.3	166.1	36.7	0.0	**2,944.7**
	Rainy days	1	2	4	10	19	20	14	16	18	12	2	0	**118**
1985	Rainfall	5.7	10.9	21.6	136.1	558.8	480.7	346.8	405.4	301.6	309.1	18.2	15.0	**2,609.9**
	Rainy days	1	3	3	6	21	21	19	19	16	23	5	2	**139**
1986	Rainfall	0.0	T	9.6	106.4	360.2	677.3	331.6	635.3	367.3	188.3	106.3	0.0	**2,782.3**
	Rainy days	0	0	1	11	17	20	17	24	20	16	5	0	**131**
1987	Rainfall	0.0	5.7	40.4	156.5	194.3	826.7	362.6	471.5	415.3	352.2	134.8	0.0	**2,960.0**
	Rainy days	0	1	5	7	8	21	18	18	20	18	13	0	**129**
1988	Rainfall	7.9	68.8	51.2	177.9	450.2	791.4	592.1	348.5	389.4	758.6	3.9	0.0	**3,639.9**
	Rainy days	2	6	4	11	21	23	15	20	20	19	2	0	**143**
1989	Rainfall	52.1	27.7	63.8	21.1	438.0	224.0	367.8	467.5	245.9	0.0	0.0	0.0	**1,907.9**
	Rainy days	4	2	4	5	18	19	13	17	22	0	0	0	**104**
1990	Rainfall	0.0	0.0	170.1	25.0	281.1	340.2	221.4	699.4	455.3	216.0	50.5	0.0	**2,459.0**
	Rainy days	0	0	5	3	16	12	10	21	16	9	4	0	**96**
Mean	**Rainfall**	**9.9**	**37.9**	**50.6**	**110.8**	**370.6**	**586.2**	**448.4**	**572.1**	**434.9**	**322.2**	**79.3**	**1.5**	**3,024.4**
Mean	**Rainy days**	**0.8**	**2.0**	**3.3**	**7.1**	**16.4**	**19.7**	**17.3**	**20.3**	**18.8**	**13.9**	**5.0**	**0.2**	**124.8**

Source: METEOROLOGICAL DEPARTMENT, 1992

Remarks: 'T' means trace rainfall with having an amount of less than 0.1 mm.

In a comparison of all study areas, it was found that the study area I has more annual rainfall and rainy days than the others. Additionally, the study area II in the Central Plain has more annual rainfall than the study area I and III but the number of rainy days is lesser. That means that it has rained somewhat more often in the study area I and III than in the Central Plain, but not heavily. Figure 18 and 19 give a better understanding of the comparison.

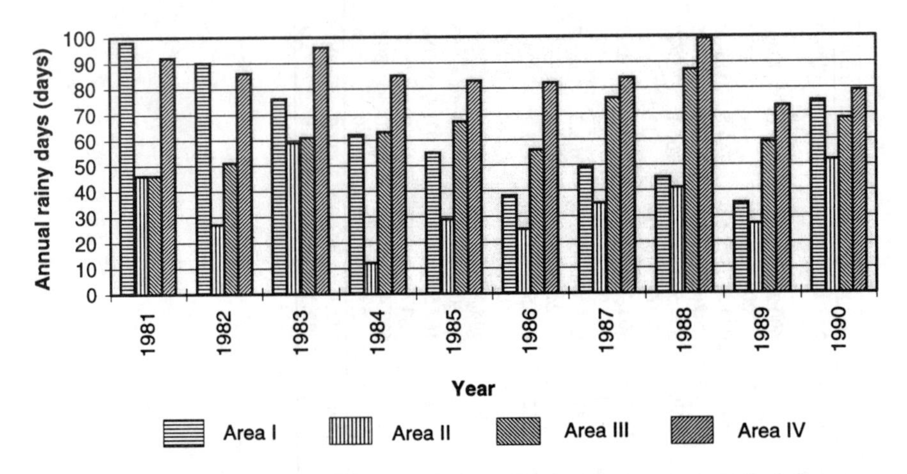

Fig. 18: **Comparison of annual rainy days of all study areas in the period of 1981-1990 (Source: Prepared by the author using the data of the Meteorological Department, Bangkok).**

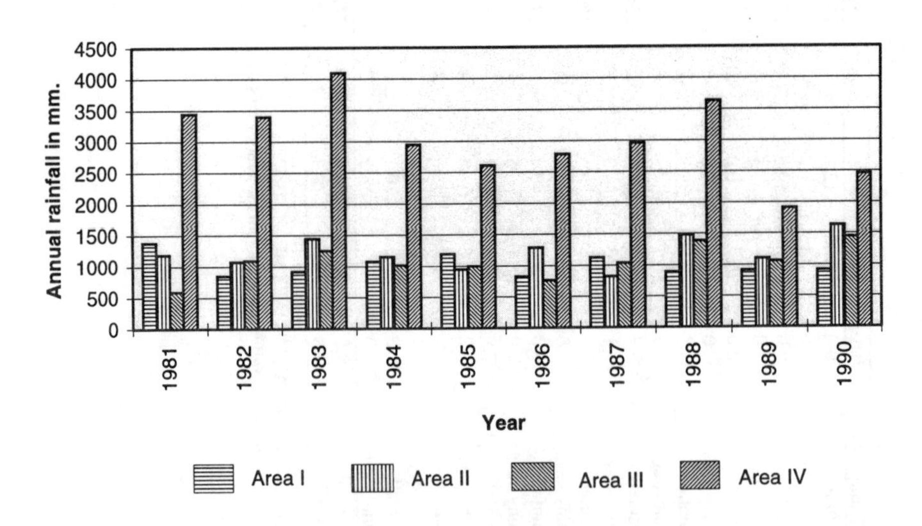

Fig. 19: **Comparison of annual rainfall (in mm) of all study areas in the period of 1981-1990 (Source: Prepared by the author using the data of the Meteorological Department, Bangkok).**

4.5.2 TEMPERATURE

Study area I (Mountainous area). The temperature data shown here are based on the available climatological data of the Meteorological Department, Bangkok. The data of this study area were recorded at the station, mentioned before, over a 10 years period (1981-1990), as shown in Tab.12. From data analysis, it can be concluded that the annual average mean temperature was 26.0° C. The annual average minimum temperature was 20.7° C, and the mean extreme minimum temperature was 7.5° C. The annual average maximum temperature was 33.2° C, and the mean extreme maximum temperature was 43.5° C. April was the warmest month, while December was the coolest month. Figure 20 shows the monthly minimum, mean, and maximum temperatures in this area.

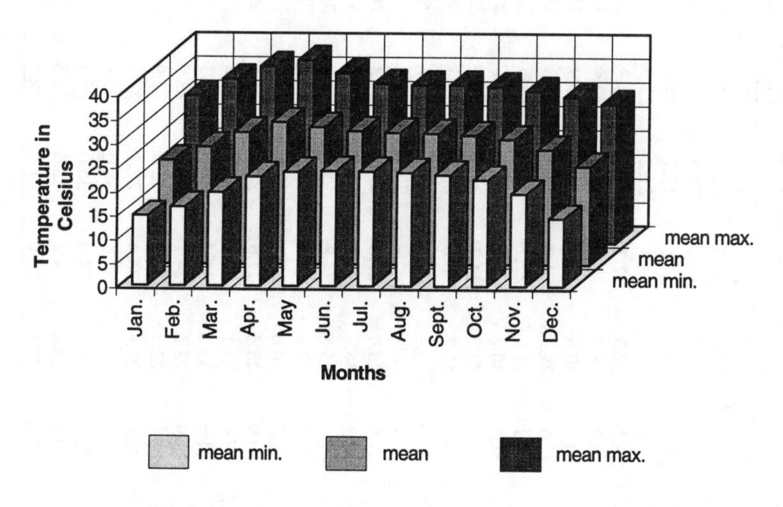

Fig. 20: Monthly minimum, mean, and maximum temperatures of the study area I (Source: Prepared by the author using the data of the Meteorological Department, Bangkok).

Study area II (Central Plain). Because no temperature data were recorded at stations situated in this study area, the temperature data shown here are based on the available climatological data of the nearest station located around 40 km the northeast of the study area. The station is located at 14° 50' latitude and 100° 31' longitude. It is about 10.0 m above MSL high. These data were recorded over a 10 years period (1981-1990), as shown in Tab. 13. From data analysis, it can be concluded that the annual average mean temperature was 28.0° C. The annual average minimum temperature was 23.7° C, and the mean extreme minimum temperature was 11.2° C. The annual average maximum temperature was 33.7° C, and the mean extreme maximum temperature was 40.5° C. April was the warmest month, while December was the coolest month. Figure 21 shows the monthly minimum, mean, and maximum temperatures in this area.

Tab. 12: **Monthly temperature over the study area I (in the mountainous area) between 1981 and 1990, measured at the Station 328003 (Location: 18° 42' Lat., 100°00' Long., 241.0 m above MSL).**

in Celsius

Year		Jan.	Feb.	Mar.	Apr.	May	Jun.	Jul.	Aug.	Sep.	Oct.	Nov.	Dec.	Annual
1981	**mean**	**20.7**	**23.8**	**27.6**	**29.5**	**28.4**	**27.6**	**27.0**	**27.2**	**27.2**	**26.3**	**24.3**	**19.9**	**25.8**
	mean max.	29.2	33.6	36.9	37.9	34.4	32.7	31.8	32.0	32.7	31.5	29.5	26.6	**32.4**
	maximum	31.6	37.3	39.2	41.6	38.3	34.4	34.6	35.2	34.9	34.2	33.1	30.7	**41.6**
	mean min.	14.2	15.5	19.5	22.5	24.1	24.3	23.7	23.8	23.4	22.7	20.5	14.7	**20.8**
	minimum	10.9	12.0	17.0	19.7	20.2	23.3	21.2	23.0	21.6	20.1	15.4	10.1	**10.1**
1982	**mean**	**20.6**	**23.5**	**27.7**	**28.5**	**29.0**	**27.7**	**27.9**	**27.3**	**26.5**	**25.9**	**24.7**	**19.6**	**25.7**
	mean max.	28.9	33.6	37.5	36.2	35.7	32.7	33.2	32.4	31.5	31.3	31.0	27.5	**32.6**
	maximum	30.8	37.2	39.5	40.4	40.6	35.0	37.1	35.9	34.6	33.7	32.1	31.4	**40.6**
	mean min.	13.8	14.8	19.3	22.5	24.2	24.2	24.2	23.8	23.2	21.9	19.4	13.0	**20.4**
	minimum	12.2	11.0	17.1	20.1	21.5	22.2	22.5	22.5	22.3	20.4	17.8	7.5	**7.5**
1983	**mean**	**20.4**	**24.7**	**28.4**	**31.9**	**30.2**	**28.8**	**28.8**	**27.6**	**27.2**	**26.4**	**22.4**	**19.7**	**26.4**
	mean max.	28.9	34.7	38.1	41.6	37.7	34.2	34.9	32.9	32.3	31.5	28.8	28.5	**33.7**
	maximum	32.4	36.5	40.6	43.5	41.2	36.6	38.6	35.5	35.7	33.7	31.5	31.6	**43.5**
	mean min.	13.5	15.9	19.7	-	-	-	24.8	24.3	23.9	23.2	18.3	13.5	-
	minimum	9.1	13.2	15.2	-	-	-	23.1	23.2	22.3	21.5	8.2	8.2	-
1984	**mean**	**21.0**	**25.6**	**27.8**	**29.9**	**28.0**	**27.6**	**27.1**	**27.2**	**26.3**	**25.3**	**23.7**	**21.9**	**26.0**
	mean max.	30.2	33.9	37.0	37.9	34.7	32.7	32.6	32.7	32.3	31.1	31.5	30.7	**33.1**
	maximum	32.5	36.9	39.6	42.2	25.5	34.8	35.7	36.3	35.7	33.6	33.7	32.7	**42.2**
	mean min.	14.0	19.6	-	23.4	23.6	24.1	23.5	23.7	22.8	21.5	18.1	15.0	-
	minimum	8.5	15.2	-	20.5	20.0	22.4	22.1	22.5	21.2	15.2	11.8	11.4	-
1985	**mean**	**23.3**	**24.9**	**28.6**	**29.9**	**28.4**	**27.6**	**27.1**	**27.4**	**27.0**	**26.0**	**24.2**	**21.2**	**26.3**
	mean max.	32.1	34.8	37.4	37.4	35.1	32.7	32.5	32.2	32.7	31.9	30.4	29.3	**33.2**
	maximum	34.2	37.4	39.7	41.6	39.2	34.7	35.2	35.3	36.2	34.4	32.8	32.0	**41.6**
	mean min.	16.0	16.6	20.4	23.9	23.6	24.1	23.4	23.9	23.2	21.9	20.0	15.2	**21.1**
	minimum	12.2	8.7	14.2	20.6	22.1	22.6	22.0	22.9	21.2	20.0	15.6	8.2	**8.2**
1986	**mean**	**20.9**	**24.2**	**25.9**	**28.8**	**27.9**	**28.0**	**27.1**	**27.5**	**26.8**	**26.1**	**24.5**	**22.0**	**25.8**
	mean max.	30.1	33.9	35.9	37.2	34.1	33.6	32.0	33.0	32.8	32.2	31.4	30.2	**33.0**
	maximum	33.5	35.7	41.7	40.1	36.7	38.0	36.0	35.2	35.2	34.3	33.8	31.7	**41.7**
	mean min.	13.6	16.0	17.1	22.6	23.9	24.1	23.6	23.8	22.8	22.1	19.7	15.9	**20.5**
	minimum	8.0	13.1	11.5	20.3	22.1	23.0	21.5	22.6	19.6	18.2	16.4	11.9	**8.0**

Tab. 12: (continued)

in Celsius

Year		Jan.	Feb.	Mar.	Apr.	May	Jun.	Jul.	Aug.	Sep.	Oct.	Nov.	Dec.	Annual
1987	**mean**	**22.2**	**23.8**	**27.2**	**29.5**	**29.9**	**28.2**	**27.8**	**27.7**	**27.3**	**26.8**	**25.7**	**19.2**	**26.3**
	mean max.	31.4	33.4	-	-	37.1	33.6	33.1	33.4	-	-	-	-	-
	maximum	33.4	36.2	-	-	41.0	37.1	36.7	37.1	-	-	-	-	-
	mean min.	15.3	16.5	19.4	22.7	24.4	24.3	24.2	23.8	23.6	22.7	21.9	12.6	**21.0**
	minimum	12.8	13.0	16.6	19.8	21.6	22.8	22.0	21.5	21.9	19.2	19.0	8.7	**8.7**
1988	**mean**	**22.2**	**25.7**	**28.6**	**29.7**	**28.7**	**27.6**	**27.8**	**27.4**	**27.4**	**26.2**	**22.5**	**20.7**	**26.2**
	mean max.	-	-	-	-	-	33.1	33.2	33.0	33.2	32.0	29.1	29.5	-
	maximum	-	-	-	-	-	35.5	35.4	35.2	35.2	34.3	32.0	32.2	-
	mean min.	14.7	17.8	20.2	23.1	24.4	23.9	24.2	24.0	23.6	22.5	17.5	14.1	**20.9**
	minimum	12.2	13.5	17.2	18.4	22.1	22.4	22.5	22.5	21.7	18.2	13.7	11.0	**11.0**
1989	**mean**	**22.7**	**24.3**	**27.4**	**30.2**	**28.4**	**27.4**	**27.7**	**27.3**	**26.9**	**26.0**	**24.0**	**19.9**	**26.0**
	mean max.	32.3	34.4	35.9	39.1	35.3	33.3	33.1	33.2	32.7	31.9	31.3	29.7	**33.5**
	maximum	35.0	37.0	39.0	41.3	39.8	35.7	36.5	34.8	35.5	34.2	33.5	33.2	**41.3**
	mean min.	15.4	15.9	20.3	22.6	23.6	23.6	24.0	23.5	23.5	22.2	18.8	12.6	**20.5**
	minimum	9.2	12.5	15.9	19.5	20.8	22.3	22.5	21.9	21.7	19.1	14.8	10.0	**9.2**
1990	**mean**	**23.0**	**24.7**	**27.1**	**29.2**	**27.9**	**28.1**	**27.5**	**27.7**	**27.0**	**26.5**	**24.6**	**21.5**	**26.2**
	mean max.	32.4	34.2	35.9	37.2	34.4	336.7	32.7	33.9	33.0	32.3	31.7	30.0	**33.5**
	maximum	34.8	36.5	38.7	40.5	39.3	36.5	35.5	37.6	34.6	34.6	34.2	33.2	**40.5**
	mean min.	15.8	17.1	20.1	22.4	23.6	24.2	23.9	23.6	23.2	22.5	19.5	15.1	**21.0**
	minimum	10.0	13.5	17.4	18.8	21.2	22.8	22.7	22.2	21.8	20.4	14.5	10.5	**10.0**
mean	mean	**21.7**	**24.5**	**27.6**	**29.7**	**28.7**	**27.9**	**27.6**	**27.4**	**27.0**	**26.2**	**24.1**	**20.6**	**26.0**
mean	mean max.	30.6	34.1	36.8	38.1	35.4	33.2	32.9	32.9	32.6	31.7	30.5	29.1	33.2
ext.	maximum	35.0	37.4	41.7	43.5	41.2	38.0	38.6	37.6	36.2	34.6	34.2	33.2	43.5
mean	mean min.	14.6	16.6	19.6	22.9	23.9	24.1	24.0	23.8	23.3	22.3	19.4	14.2	20.7
ext.	minimum	8.0	8.7	11.5	18.4	20.0	22.2	21.2	21.5	19.6	15.2	8.2	7.5	7.5

Source: METEOROLOGICAL DEPARTMENT, 1992
Remarks: '-' means missing data

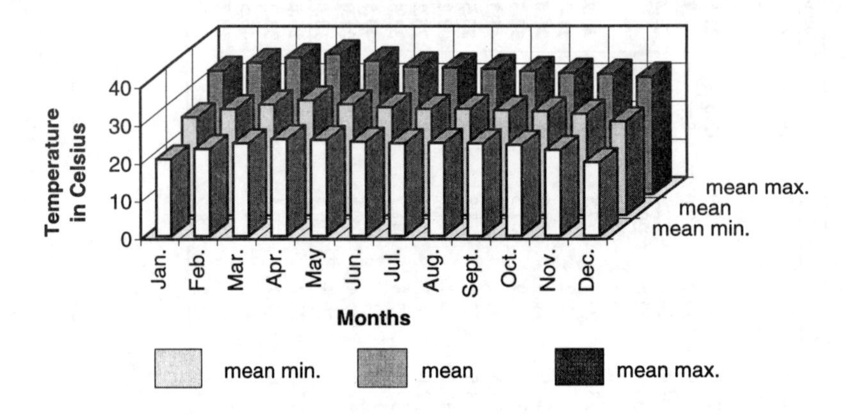

Fig. 21: **Monthly minimum, mean, and maximum temperatures of the study area II (Source: Prepared by the author using the data of the Meteorological Department, Bangkok).**

Study area III (Korat plateau). The temperature data shown here are also based on the available climatological data of the Meteorological Department at the station mentioned before. The data were recorded over a 10 years period (1981-1990), shown in Tab. 14. From the data analysis, it can be concluded that the annual average mean temperature was 27.0° C. The annual average minimum temperature was 22.1° C, and the mean extreme minimum temperature was 7.9° C. The annual average maximum temperature was 32.7° C, and the mean extreme maximum temperature was 42.6° C. April was the warmest month, while December was the coolest month. Figure 22 shows the monthly minimum, mean, and maximum temperatures in this area.

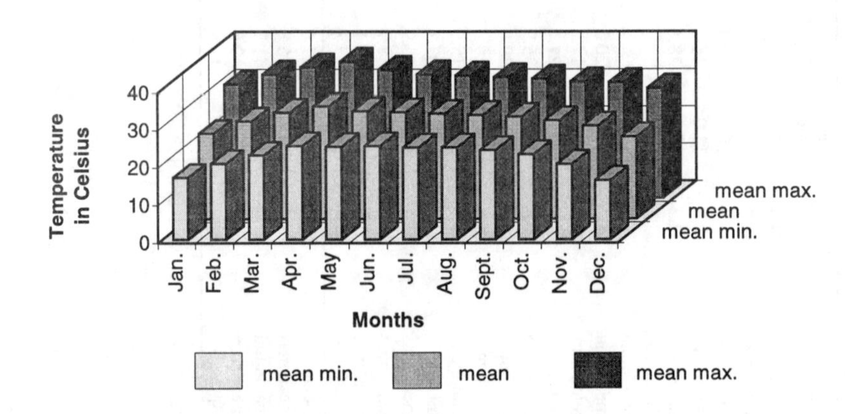

Fig. 22: **Monthly minimum, mean, and maximum temperatures of the study area III (Source: Prepared by the author using the data of the Meteorological Department, Bangkok).**

Tab. 13: **Monthly temperature over the study area II (in the Central Plain) between 1981 and 1990, measured at the Station 426002 (Location: 14° 50' Lat., 100°31' Long., 10.0 m above MSL).**

in Celsius

Year		Jan.	Feb.	Mar.	Apr.	May	Jun.	Jul.	Aug.	Sep.	Oct.	Nov.	Dec.	Annual
1981	**mean**	**25.1**	**27.8**	**29.4**	**29.8**	**28.8**	**28.2**	**28.0**	**27.8**	**27.8**	**27.9**	**26.4**	**24.0**	**27.6**
	mean max.	31.9	34.8	35.8	36.2	34.4	32.6	32.2	32.1	32.5	31.9	30.7	29.9	**32.9**
	maximum	36.0	37.2	37.5	38.0	37.2	34.4	34.5	34.0	34.5	33.7	33.2	32.6	**38.0**
	mean min.	19.0	22.3	24.6	25.1	25.2	24.8	24.8	24.8	24.6	24.6	23.3	18.6	**23.5**
	minimum	14.0	16.4	21.8	21.9	22.0	23.2	23.0	22.7	22.2	23.3	19.0	12.5	**12.5**
1982	**mean**	**25.2**	**28.2**	**29.6**	**29.3**	**29.7**	**28.3**	**28.1**	**27.6**	**27.3**	**27.6**	**28.0**	**23.5**	**27.7**
	mean max.	32.4	34.5	36.2	35.2	35.5	33.2	33.0	31.9	31.3	32.2	32.8	29.5	**33.1**
	maximum	34.0	37.0	38.7	38.0	38.4	35.0	34.5	35.2	34.0	33.7	34.2	33.0	**38.7**
	mean min.	18.2	23.2	25.2	25.2	25.8	25.1	24.4	24.6	24.5	24.0	23.7	17.8	**23.5**
	minimum	14.0	19.5	23.3	23.2	22.4	23.5	23.2	23.6	22.5	21.6	21.5	11.2	**11.2**
1983	**mean**	**25.32**	**28.3**	**29.7**	**31.5**	**30.5**	**29.0**	**28.8**	**28.2**	**27.9**	**26.9**	**25.3**	**24.8**	**28.0**
	mean max.	31.6	34.9	36.4	38.2	37.0	33.7	33.9	32.6	32.3	31.0	30.2	30.5	**33.5**
	maximum	34.2	36.2	39.0	40.3	39.8	36.5	36.7	34.7	33.6	33.3	33.3	34.1	**40.3**
	mean min.	19.7	23.3	24.9	27.0	26.0	25.4	24.9	24.8	24.8	24.2	21.3	19.8	**23.9**
	minimum	13.7	20.6	22.5	25.0	23.2	23.0	23.3	22.8	22.8	22.6	13.8	13.9	**13.7**
1984	**mean**	**25.0**	**27.8**	**28.6**	**29.9**	**29.6**	**28.5**	**28.0**	**27.9**	**27.4**	**26.9**	**26.7**	**25.8**	**27.7**
	mean max.	31.2	33.7	35.0	36.1	34.7	34.0	33.4	33.2	32.5	32.0	32.4	32.1	**33.4**
	maximum	34.8	35.8	37.8	38.2	37.7	36.7	36.0	35.2	35.0	33.5	35.2	34.7	**38.2**
	mean min.	19.8	23.2	24.0	25.2	-	25.0	24.4	24.5	24.3	23.3	22.4	20.9	-
	minimum	12.0	18.5	21.6	21.9	-	24.0	22.3	22.7	22.2	19.0	19.2	16.7	-
1985	**mean**	**26.6**	**28.7**	**30.0**	**30.4**	**29.0**	**28.0**	**27.5**	**28.0**	**27.3**	**27.0**	**27.1**	**24.8**	**27.9**
	mean max.	33.5	35.8	37.0	37.3	35.1	33.1	32.8	33.2	32.2	32.2	32.4	31.8	**33.9**
	maximum	35.4	37.6	39.4	40.5	38.7	35.5	34.8	34.8	34.6	34.0	34.1	34.7	**40.5**
	mean min.	21.1	23.9	25.0	25.9	25.4	24.9	24.2	24.8	24.4	23.8	23.2	19.2	**23.9**
	minimum	17.5	20.9	19.2	23.0	23.0	23.0	22.7	23.5	23.2	21.7	17.6	14.2	**14.2**
1986	**mean**	**24.9**	**27.5**	**28.6**	**29.7**	**28.7**	**28.8**	**27.5**	**28.2**	**28.3**	**27.6**	**26.8**	**25.7**	**27.7**
	mean max.	32.1	34.4	36.0	36.8	34.1	34.6	32.9	33.4	33.3	32.5	32.3	31.9	**33.7**
	maximum	35.8	36.6	39.0	39.0	36.2	36.9	37.2	35.5	35.5	35.2	33.8	34.8	**39.0**
	mean min.	19.0	22.5	23.7	25.4	25.3	24.9	24.4	24.6	24.7	24.5	22.3	20.6	**23.5**
	minimum	15.9	19.3	16.4	23.1	23.3	23.0	23.0	23.6	23.0	22.8	18.9	17.2	**15.9**

Tab. 13: (continued)

in Celsius

Year		Jan.	Feb.	Mar.	Apr.	May	Jun.	Jul.	Aug.	Sep.	Oct.	Nov.	Dec.	Annual
1987	**mean**	**26.2**	**27.9**	**29.2**	**30.7**	**30.1**	**29.2**	**29.1**	**28.9**	**28.0**	**28.1**	**27.7**	**23.6**	**28.2**
	mean max.	33.0	34.7	36.4	37.9	37.0	34.7	35.1	34.4	32.8	33.1	32.3	30.1	**34.4**
	maximum	35.2	37.2	38.7	40.5	38.9	36.2	37.7	37.5	35.5	34.7	34.5	34.1	**40.5**
	mean min.	20.9	23.1	24.3	25.9	25.9	25.1	24.4	24.6	24.5	24.6	24.5	18.2	**23.9**
	minimum	16.7	17.5	20.3	23.1	23.4	23.4	23.0	23.1	23.0	21.1	22.1	13.9	**13.9**
1988	**mean**	**26.7**	**28.5**	**30.3**	**30.1**	**29.0**	**28.7**	**28.4**	**28.0**	**28.3**	**27.2**	**25.7**	**25.2**	**28.0**
	mean max.	33.9	34.9	37.1	36.9	34.6	34.2	33.9	33.3	33.5	31.7	30.6	30.8	**33.8**
	maximum	36.6	37.5	39.2	39.4	36.6	36.2	36.0	34.4	35.4	33.8	32.6	33.2	**39.4**
	mean min.	20.7	-	25.6	25.8	25.5	25.0	25.0	24.6	25.1	24.3	22.1	20.3	**-**
	minimum	15.7	-	22.2	22.7	24.0	23.4	23.5	23.0	23.2	21.3	19.2	15.2	**-**
1989	**mean**	**27.3**	**27.7**	**28.6**	**31.0**	**29.5**	**28.4**	**28.5**	**28.2**	**27.7**	**27.7**	**26.9**	**24.9**	**28.0**
	mean max.	33.3	34.2	35.00	37.8	35.5	33.7	34.2	33.7	32.7	32.3	32.1	31.6	**33.8**
	maximum	35.5	36.1	37.5	39.3	38.8	35.4	36.8	35.5	34.4	33.9	34.0	33.9	**39.3**
	mean min.	22.6	22.9	24.1	25.8	25.1	24.2	24.7	24.4	23.8	23.9	22.6	19.0	**23.6**
	minimum	18.7	19.9	20.5	23.5	23.3	22.5	22.9	23.4	22.1	20.3	18.0	14.2	**14.2**
1990	**mean**	**27.6**	**28.7**	**29.4**	**31.1**	**29.4**	**29.3**	**28.5**	**28.8**	**27.9**	**27.6**	**27.0**	**26.1**	**28.5**
	mean max.	24.0	25.6	26.3	28.1	25.5	24.5	24.0	24.2	32.9	31.9	31.8	32.0	**34.2**
	maximum	35.3	37.1	38.8	40.5	39.2	36.0	36.3	36.7	35.8	34.4	34.0	33.9	**40.5**
	mean min.	22.5	23.5	24.5	25.6	25.0	25.1	24.5	24.8	24.0	24.2	22.6	20.5	**23.9**
	minimum	19.0	21.9	21.6	22.6	22.5	23.0	23.3	22.8	22.4	22.2	18.1	17.1	**17.1**
mean	mean	**26.0**	**28.1**	**29.3**	**30.4**	**29.4**	**28.6**	**28.2**	**28.2**	**27.8**	**27.5**	**26.8**	**24.8**	**28.0**
mean	mean max.	32.7	34.8	36.1	37.1	35.3	33.8	33.5	33.2	32.6	32.1	31.8	31.0	33.7
ext.	maximum	36.6	37.6	39.4	40.5	39.8	36.9	37.7	37.5	35.8	35.2	35.2	34.8	40.5
mean	mean min.	20.4	23.1	24.6	25.7	25.5	25.0	24.6	24.7	24.5	24.1	22.8	19.5	23.7
ext.	minimum	12.0	16.4	16.4	21.9	22.0	22.5	22.3	22.7	22.1	19.0	13.8	11.2	11.2

Source: METEOROLOGICAL DEPARTMENT, 1992

Remarks: '-' means missing data

Tab. 14: **Monthly temperature over the study area III (in the Korat plateau) between 1981 and 1990, measured at the Station 381003 (Location: 16° 29' Lat., 102°07' Long., 170.0 m above MSL).**

in Celsius

Year		Jan.	Feb.	Mar.	Apr.	May	Jun.	Jul.	Aug.	Sep.	Oct.	Nov.	Dec.	Annual
1981	**mean**	**22.0**	**26.5**	**29.5**	**29.6**	**28.7**	**28.2**	**27.4**	**27.8**	**27.8**	**26.4**	**25.2**	**21.0**	**26.7**
	mean max.	30.2	34.0	36.8	36.4	34.4	33.4	31.8	32.7	33.1	31.5	31.1	28.3	**32.8**
	maximum	33.2	38.2	40.0	40.1	37.5	35.6	35.1	35.5	35.9	35.3	34.5	33.1	**40.1**
	mean min.	15.2	20.2	23.1	24.5	24.8	24.7	24.1	24.4	24.0	22.8	20.8	15.1	**22.0**
	minimum	10.1	15.9	20.1	21.5	21.8	23.5	21.5	22.0	22.2	19.2	14.8	10.4	**10.1**
	mean	**22.4**	**25.8**	**29.3**	**28.0**	**29.7**	**29.2**	**28.4**	**27.6**	**26.5**	**26.7**	**25.9**	**20.3**	**26.7**
	mean max.	30.7	32.6	36.1	34.0	35.8	34.3	33.7	31.9	31.3	31.9	32.3	28.2	**32.7**
	maximum	33.4	37.5	39.4	39.5	39.5	37.4	36.1	35.9	34.6	33.6	34.0	33.5	**39.5**
	mean min.	15.5	20.1	23.8	23.2	24.8	25.0	24.7	24.3	23.4	22.8	21.2	14.1	**21.9**
	minimum	12.2	15.9	21.4	18.9	21.7	23.1	22.5	22.5	22.0	20.8	16.6	7.9	**7.9**
1983	**mean**	**21.5**	**26.8**	**28.7**	**32.5**	**30.0**	**29.1**	**28.8**	**27.5**	**27.8**	**26.5**	**23.1**	**21.3**	**27.0**
	mean max.	29.2	34.7	36.2	40.1	36.4	33.9	34.0	31.9	32.8	31.1	30.1	29.2	**33.3**
	maximum	33.8	36.6	40.6	42.6	41.1	37.5	37.0	34.2	34.7	34.0	33.1	33.0	**42.6**
	mean min.	15.3	20.6	-	26.3	25.5	25.3	25.0	24.5	24.3	23.3	18.0	15.3	**-**
	minimum	8.2	16.0	-	23.7	23.0	21.9	22.9	22.7	22.5	21.0	10.3	9.5	
1984	**mean**	**21.4**	**26.1**	**28.8**	**31.0**	**28.0**	**28.1**	**27.5**	**27.6**	**26.8**	**25.6**	**24.6**	**22.6**	**26.5**
	mean max.	29.8	33.4	36.4	38.0	33.8	32.7	32.3	31.8	31.3	30.4	30.5	30.1	**32.5**
	maximum	34.4	38.4	41.1	41.0	36.8	.5.5	34.9	35.0	34.3	32.9	33.1	33.0	**41.1**
	mean min.	14.7	20.2	22.2	25.8	23.9	24.9	23.9	24.4	23.4	21.8	20.0	16.5	**21.8**
	minimum	9.0	13.5	14.4	23.6	21.9	22.7	21.8	22.6	22.1	16.0	16.5	13.1	**9.0**
1985	**mean**	**23.0**	**26.7**	**27.6**	**29.7**	**27.8**	**28.1**	**27.6**	**27.8**	**26.9**	**25.2**	**25.8**	**22.2**	**26.5**
	mean max.	30.4	33.9	34.4	35.8	33.6	32.2	32.4	32.1	31.3	30.8	31.6	30.0	**32.4**
	maximum	33.9	37.6	39.0	41.2	39.3	35.3	34.7	35.4	34.7	33.6	34.1	33.1	**41.2**
	mean min.	17.0	20.8	21.6	24.6	24.6	24.8	24.0	24.5	23.4	22.2	21.0	16.0	**22.1**
	minimum	13.8	15.1	15.7	20.2	22.6	22.4	21.6	22.1	21.7	18.5	18.9	10.4	**10.4**
1986	**mean**	**21.5**	**25.5**	**-**	**29.6**	**27.8**	**28.7**	**28.0**	**27.9**	**27.3**	**26.6**	**24.2**	**22.9**	**-**
	mean max.	29.5	32.9	35.1	36.0	32.4	33.5	32.6	32.6	32.2	31.7	30.4	30.1	**32.4**
	maximum	34.1	37.2	40.2	39.5	35.5	36.3	36.6	35.6	34.5	34.1	33.0	33.2	**40.2**
	mean min.	14.8	19.2	20.9	24.3	24.4	25.0	24.5	24.5	23.3	22.7	19.1	17.1	**21.7**
	minimum	10.1	15.4	11.1	20.0	21.9	22.1	22.1	22.4	20.3	19.7	15.6	13.9	**10.1**

Tab. 14: (continued)

in Celsius

Year		Jan.	Feb.	Mar.	Apr.	May	Jun.	Jul.	Aug.	Sep.	Oct.	Nov.	Dec.	Annual
1987	**mean**	**23.5**	**26.1**	**28.5**	**29.6**	**29.7**	**28.9**	**28.9**	**28.0**	**27.1**	**26.9**	**26.4**	**20.5**	**27.0**
	mean max.	31.1	33.1	35.1	35.6	35.7	33.9	33.8	32.7	31.2	32.0	31.5	28.1	**32.8**
	maximum	34.9	37.2	39.6	39.4	38.4	36.1	36.9	35.5	33.8	34.6	33.6	32.4	**39.6**
	mean min.	17.1	20.3	22.7	24.5	25.2	25.1	25.4	24.4	24.1	23.3	22.4	14.2	**22.4**
	minimum	13.9	14.4	17.9	21.3	21.9	22.2	23.6	21.9	21.8	17.7	16.0	9.9	**9.9**
1988	**mean**	**24.1**	**26.4**	**29.5**	**29.8**	**28.7**	**28.2**	**28.2**	**28.1**	**27.7**	**25.9**	**23.2**	**22.3**	**26.3**
	mean max.	32.0	33.0	35.9	35.4	33.5	32.9	33.3	32.6	32.8	30.3	29.5	29.6	**32.6**
	maximum	36.2	36.6	40.6	40.2	38.0	35.8	35.5	35.0	36.3	34.0	32.0	31.6	**40.6**
	mean min.	17.6	21.3	24.0	25.3	25.0	24.8	24.4	24.7	23.8	22.7	18.1	16.3	**22.4**
	minimum	12.9	15.0	16.8	21.7	22.8	22.5	21.0	23.5	22.0	18.8	14.8	12.5	**12.5**
1989	**mean**	**23.9**	**25.1**	**26.5**	**30.7**	**29.0**	**28.0**	**28.0**	**27.5**	**27.4**	**26.2**	**24.9**	**22.6**	**26.7**
	mean max.	30.8	32.3	32.5	37.0	34.3	32.6	32.6	32.3	32.2	31.0	31.3	30.6	**32.5**
	maximum	35.4	37.4	38.0	40.9	38.0	35.2	35.6	35.2	34.0	34.5	33.5	33.4	**40.9**
	mean min.	18.3	19.1	21.4	25.2	24.8	24.3	24.5	24.3	23.9	22.1	19.8	16.1	**22.0**
	minimum	14.4	14.0	15.1	22.0	22.5	21.7	22.1	22.4	22.5	18.7	14.7	12.7	**12.7**
1990	**mean**	**24.7**	**25.6**	**26.8**	**30.5**	**28.5**	**28.3**	**27.8**	**27.9**	**27.4**	**26.5**	**25.3**	**23.0**	**26.9**
	mean max.	31.7	32.4	32.8	37.2	34.0	33.0	32.5	33.2	32.2	31.4	31.3	30.4	**32.7**
	maximum	34.9	35.8	37.9	41.1	39.8	35.0	35.2	35.5	34.2	34.1	34.6	33.8	**41.1**
	mean min.	19.2	20.1	21.9	25.3	24.3	25.1	24.6	24.5	23.9	22.9	20.5	17.0	**22.5**
	minimum	15.9	16.0	19.1	21.3	22.0	23.2	23.1	22.7	20.8	20.7	16.2	12.7	**12.7**
mean	mean	**22.8**	**26.1**	**28.4**	**30.1**	**28.8**	**28.5**	**28.1**	**27.8**	**27.3**	**26.3**	**24.9**	**21.9**	**27.0**
mean	mean max.	30.5	33.2	35.1	36.6	34.4	33.2	32.9	32.4	32.0	31.2	31.0	29.5	32.7
ext.	maximum	36.2	38.4	41.1	42.6	41.1	37.5	37.0	35.9	36.3	35.3	34.6	33.8	42.6
mean	mean min.	16.5	20.2	22.4	24.9	24.7	24.9	24.5	24.5	23.8	22.7	20.1	15.8	22.1
ext.	minimum	8.2	13.5	11.1	18.9	21.7	21.7	21.0	21.9	20.3	16.0	10.3	7.9	7.9

Source: METEOROLOGICAL DEPARTMENT, 1992
Remarks: '-' means missing data

Tab. 15: **Monthly temperature over the study area IV (in the coastal zone) between 1981 and 1990, measured at the Station 480003 (Location: 12° 22' Lat., 102°21' Long., 3.0 m above MSL).**

in Celsius

Year		Jan.	Feb.	Mar.	Apr.	May	Jun.	Jul.	Aug.	Sep.	Oct.	Nov.	Dec.	Annual
1981	**mean**	**25.5**	**26.8**	**27.6**	**27.8**	**27.7**	**27.4**	**27.1**	**27.2**	**26.9**	**27.1**	**26.6**	**24.6**	**26.9**
	mean max.	33.3	32.6	32.4	32.5	32.0	30.4	30.3	30.1	31.9	31.5	31.7	30.9	**31.6**
	maximum	35.5	34.7	34.2	35.2	33.5	32.2	31.7	31.8	34.3	35.5	34.9	35.0	**35.5**
	mean min.	19.5	22.1	23.5	24.4	24.7	25.0	24.1	24.7	23.5	23.5	22.5	19.0	**23.1**
	minimum	16.2	16.8	21.1	21.9	23.1	23.6	22.2	22.6	21.2	22.2	18.5	15.0	**15.0**
1982	**mean**	**25.5**	**27.2**	**27.9**	**27.6**	**28.1**	**27.2**	**27.2**	**26.7**	**26.7**	**27.2**	**27.9**	**25.3**	**27.0**
	mean max.	34.1	32.5	32.2	32.8	32.2	30.6	30.0	29.7	30.0	32.8	34.4	32.5	**32.0**
	maximum	35.7	35.8	34.3	35.2	34.3	32.1	31.6	31.1	33.0	35.6	36.0	35.0	**36.0**
	mean min.	-	22.8	23.6	23.1	25.2	24.7	24.8	24.8	24.4	23.6	23.8	20.3	-
	minimum	-	20.5	21.6	21.4	23.4	22.8	22.8	22.8	23.0	22.2	22.9	15.4	-
1983	**mean**	**26.4**	**27.4**	**28.0**	**29.4**	**28.1**	**27.8**	**27.3**	**27.1**	**27.1**	**26.5**	**25.6**	**25.5**	**27.2**
	mean max.	33.3	32.6	33.0	34.2	32.0	30.9	30.3	30.4	31.1	30.8	30.6	30.9	**31.7**
	maximum	36.3	33.6	34.5	35.0	34.5	33.1	32.0	32.7	32.9	34.2	34.4	33.4	**36.3**
	mean min.	21.2	23.2	24.0	25.6	25.1	25.4	25.0	24.8	24.5	24.0	22.4	21.6	**23.9**
	minimum	15.2	19.0	21.9	24.0	23.3	22.9	23.0	23.0	23.5	22.5	17.7	18.6	**15.2**
1984	**mean**	**25.4**	**27.3**	**27.7**	**28.6**	**27.7**	**27.5**	**27.6**	**27.7**	**26.6**	**26.7**	**26.8**	**26.3**	**27.2**
	mean max.	31.4	32.9	33.3	33.5	32.0	30.7	31.3	30.6	31.1	31.7	32.5	32.4	**32.0**
	maximum	34.4	34.8	35.4	35.2	34.6	33.8	33.2	35.5	33.2	34.1	35.2	34.4	**35.5**
	mean min.	20.8	23.4	23.4	25.0	24.9	25.3	24.8	25.4	24.0	23.1	22.8	21.6	**23.7**
	minimum	16.6	19.7	21.5	23.5	22.9	23.5	23.5	22.6	22.3	19.6	20.5	19.1	**16.6**
1985	**mean**	**26.2**	**27.6**	**28.1**	**28.1**	**27.5**	**27.5**	**26.9**	**27.5**	**27.0**	**26.5**	**27.2**	**25.6**	**27.1**
	mean max.	32.9	32.7	33.4	33.1	32.0	30.6	30.6	30.7	31.1	31.4	32.2	31.6	**31.9**
	maximum	34.6	34.1	35.6	35.3	34.1	32.8	32.4	31.8	33.5	34.6	34.3	34.4	**35.6**
	mean min.	20.9	23.7	24.1	24.8	24.4	25.1	24.3	25.2	24.5	23.7	23.7	20.6	**23.8**
	minimum	17.8	20.0	21.8	22.9	22.3	22.3	23.1	23.0	22.5	22.5	21.3	16.3	**16.3**
1986	**mean**	**24.9**	**26.4**	**27.5**	**28.3**	**28.0**	**27.6**	**27.5**	**27.3**	**27.0**	**26.9**	**26.5**	**25.6**	**27.0**
	mean max.	31.3	32.1	32.8	33.7	31.5	31.2	31.2	30.5	31.4	32.1	31.4	31.3	**31.7**
	maximum	33.8	35.0	35.2	35.6	34.7	33.0	33.0	31.8	33.5	34.1	33.5	33.8	**35.6**
	mean min.	19.9	21.6	23.2	24.6	25.2	25.0	24.8	24.9	24.3	24.1	23.3	20.9	**23.5**
	minimum	16.7	19.5	14.5	22.8	23.3	23.4	22.6	22.5	22.8	23.3	20.5	17.5	**14.5**

Tab. 15: (continued)

in Celsius

Year		Jan.	Feb.	Mar.	Apr.	May	Jun.	Jul.	Aug.	Sep.	Oct.	Nov.	Dec.	Annual
1987	**mean**	**26.3**	**26.8**	**27.8**	**28.7**	**28.6**	**28.3**	**28.1**	**27.5**	**27.1**	**27.3**	**27.3**	**24.3**	**27.3**
	mean max.	32.7	32.4	32.7	33.8	33.4	32.0	31.8	31.7	31.6	32.8	32.0	29.4	**32.2**
	maximum	35.5	34.5	34.7	35.3	35.2	34.5	33.3	34.2	34.5	35.0	34.0	33.9	**35.5**
	mean min.	21.0	22.2	23.7	24.8	25.3	25.6	25.4	24.6	24.3	23.9	24.3	20.1	**23.8**
	minimum	18.1	17.4	21.6	23.1	23.7	23.3	22.7	23.2	22.8	22.5	21.8	16.1	**16.1**
1988	**mean**	**26.7**	**27.6**	**28.7**	**28.6**	**28.3**	**27.6**	**27.6**	**27.4**	**27.4**	**26.3**	**25.9**	**25.2**	**27.3**
	mean max.	32.9	32.8	33.4	33.9	32.4	31.4	31.5	31.5	32.0	30.6	30.0	31.0	**32.0**
	maximum	34.6	35.5	34.9	35.0	34.2	33.8	34.6	33.1	34.1	33.4	33.7	33.7	**35.5**
	mean min.	21.6	23.5	25.1	25.1	25.4	24.8	24.7	24.7	24.7	23.7	22.7	20.4	**23.9**
	minimum	18.4	22.0	23.0	23.8	23.8	23.2	23.4	23.0	23.9	21.5	19.7	17.2	**17.2**
1989	**mean**	**27.1**	**26.6**	**27.5**	**28.9**	**28.2**	**27.6**	**27.6**	**27.4**	**27.1**	**26.8**	**26.6**	**25.3**	**27.2**
	mean max.	32.4	31.8	32.9	33.7	32.6	31.5	31.3	31.1	31.3	31.9	31.8	31.5	**32.0**
	maximum	34.6	33.9	34.2	35.0	34.9	34.1	33.2	32.5	34.5	34.2	33.6	34.7	**35.0**
	mean min.	22.8	22.3	23.2	24.9	25.1	24.9	24.8	24.9	24.4	23.9	22.7	20.2	**23.7**
	minimum	20.0	18.9	21.4	22.7	23.4	22.5	23.2	23.0	22.3	22.4	19.2	17.4	**17.4**
1990	**mean**	**26.7**	**27.6**	**27.4**	**29.3**	**28.3**	**28.5**	**27.9**	**27.6**	**27.3**	**27.1**	**26.5**	**26.1**	**27.5**
	mean max.	32.6	32.9	32.6	34.4	33.0	31.8	31.7	31.3	31.4	31.9	31.6	32.1	**32.3**
	maximum	34.0	34.1	34.2	35.8	35.4	33.5	33.9	32.9	33.0	33.9	34.0	34.5	**35.8**
	mean min.	22.0	23.3	23.4	25.2	25.1	26.0	25.3	25.1	24.7	24.3	22.8	21.3	**24.1**
	minimum	19.8	20.7	22.1	23.0	23.2	24.0	23.0	22.6	22.3	22.6	18.8	17.9	**17.9**
mean	mean	**26.1**	**27.1**	**27.8**	**28.5**	**28.1**	**27.7**	**27.5**	**27.3**	**27.0**	**26.8**	**26.7**	**25.4**	**27.0**
mean	mean max.	32.7	32.5	32.9	33.6	32.3	31.1	31.0	30.8	31.3	31.8	31.8	31.4	31.9
ext.	maximum	36.3	35.8	35.6	35.8	35.4	34.5	34.6	35.5	34.5	35.6	36.0	35.0	36.3
mean	mean min.	21.1	22.8	23.7	24.8	25.0	25.2	24.8	24.9	24.3	23.8	23.1	20.6	23.7
ext.	minimum	15.2	16.8	14.5	21.4	22.3	22.3	22.2	22.5	21.2	19.6	17.7	15.0	14.5

Source: METEOROLOGICAL DEPARTMENT, 1992

Remarks: '-' means missing data

Study area IV (Coastal zone). The temperature data showed here are also based on the available climatological data of the Meteorological Department. The data were also recorded at the station 48003 over a 10 years period (1981-1990), shown in Tab. 15. From data analysis, it can be concluded that the annual average mean temperature was 27.0° C. The annual average minimum temperature was 23.7° C, and the mean extreme minimum temperature was 14.5° C. The annual average maximum temperature was 31.9° C, and the mean extreme maximum temperature was 36.3 °C. April was the hottest month, while December was the coolest month. Figure 23 shows the monthly minimum, mean, and maximum temperatures in this area.

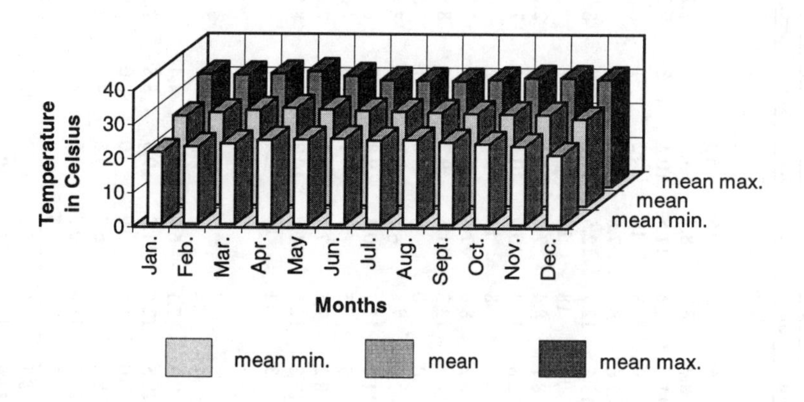

Fig. 23: **Monthly minimum, mean, and maximum temperatures of the study area IV (Source: Prepared by the author using the data of the Meteorological Department, Bangkok).**

It can be seen that the temperature of all study areas behave in the same manner. April is the warmest month and the December is the coolest month for all study areas. However, the temperature of the study area I and III are more extreme than the temperature of the study area II and IV.

4.5.3 CLOUDINESS, VISIBILITY AND THUNDERSTORM DATA

Table 16 to 19 show more climatological data of the study areas, especially the cloudiness, visibility, and number of thunderstorm days. These data are normally essential for aerial photography. Cloudiness data can also help us to select a proper date of satellite data used. Visibility and thunderstorm data could be also essential for field work. All data are derived from the Meterological Department and from the stations as the other climatological data mentioned before. The cloudiness data are measured and assigned with the value of 0 to 10. The visibility data are measured in kilometer. The thunderstorm data are noted as the number of days. Figures 24, 25, 26, and 27 show the monthly measured values of the data in each study areas.

Tab. 16: Cloudiness, visibility, and thunderstorm data of the study area I (in the mountainous area) between 1981 and 1990, measured at the Station 328003 (Location : 18° 42' Lat., 100°00' Long., 241.0 m above MSL).

Year		Jan.	Feb.	Mar.	Apr.	May	Jun.	Jul.	Aug.	Sep.	Oct.	Nov.	Dec.	**Annual**
1981	Mean Cloudiness (0-10)	-	-	-	-	-	-	-	-	-	-	-	-	**-**
	Mean Visibility (km.)	-	-	-	-	-	-	-	-	-	-	-	-	**-**
	Thunderstorm (days)	0	0	9	15	20	6	6	7	15	10	4	0	**92**
1982	Cloudiness (0-10)	1.8	0.8	1.6	5.4	6.0	8.2	8.7	8.9	8.2	5.6	3.8	2.2	**5.1**
	Visibility (km.)	6.0	5.5	3.9	7.2	10.2	13.1	11.1	12.6	11.5	10.4	11.8	9.2	**9.4**
	Thunderstorm (days)	0	0	2	13	14	5	9	3	18	12	1	0	**77**
1983	Cloudiness (0-10)	3.1	1.5	0.8	1.8	6.0	7.2	7.6	8.4	7.8	7.7	5.4	-	**-**
	Visibility (km.)	6.7	4.7	4.0	4.2	10.2	13.8	13.5	12.7	12.5	11.3	9.7	8.3	**9.3**
	Thunderstorm (days)	1	0	0	3	16	6	14	18	16	8	1	0	**83**
1984	Cloudiness (0-10)	2.3	4.2	1.7	4.2	6.6	8.5	8.1	8.4	7.8	6.6	3.6	2.7	**5.4**
	Visibility (km.)	7.8	5.5	4.2	6.7	10.8	12.7	12.8	12.3	10.9	10.4	8.6	7.4	**9.2**
	Thunderstorm (days)	0	6	1	8	15	11	12	8	14	8	2	0	**85**
1985	Cloudiness (0-10)	2.4	2.9	2.1	4.5	6.7	8.4	8.2	8.8	7.4	6.5	-	3.3	**-**
	Visibility (km.)	5.1	4.6	3.8	7.1	11.7	12.3	12.2	12.3	11.5	9.7	9.5	8.5	**9.0**
	Thunderstorm (days)	0	1	0	1	0	0	0	0	0	0	1	0	**3**
1986	Cloudiness (0-10)	2.2	1.4	1.7	4.3	7.1	7.5	7.9	7.6	6.4	5.9	5.4	3.5	**5.1**
	Visibility (km.)	6.4	3.5	3.6	5.6	11.0	12.5	11.7	11.7	10.4	9.8	9.5	8.1	**8.7**
	Thunderstorm (days)	0	0	0	19	13	8	5	15	17	7	0	0	**83**
1987	Cloudiness (0-10)	2.0	2.9	3.2	4.3	5.2	7.2	8.2	7.5	7.1	5.1	6.3	2.6	**5.1**
	Visibility (km.)	6.2	3.6	7.0	7.1	12.7	12.8	13.3	12.7	11.9	10.3	10.6	8.5	**9.7**
	Thunderstorm (days)	0	2	4	11	18	14	3	11	12	14	7	0	**96**
1988	Cloudiness (0-10)	-	-	-	-	-	-	-	-	-	-	-	-	**-**
	Visibility (km.)	-	-	-	-	-	-	-	-	-	-	-	-	**-**
	Thunderstorm (days)	-	-	-	-	-	-	-	-	-	-	-	-	**-**
1989	Cloudiness (0-10)	2.7	1.2	3.6	1.5	6.1	7.2	7.3	7.1	6.8	6.0	4.4	2.1	**4.7**
	Visibility (km.)	7.9	6.6	8.0	8.5	11.8	12.8	12.5	12.9	11.0	9.5	8.8	7.5	**9.8**
	Thunderstorm (days)	0	0	4	6	20	11	10	15	19	10	0	0	**95**
1990	Cloudiness (0-10)	2.3	2.9	2.9	3.9	7.2	7.7	8.8	6.9	6.9	5.6	4.1	2.5	**5.1**
	Visibility (km.)	6.0	5.7	7.4	6.7	9.8	11.5	11.7	11.8	10.5	8.5	8.2	5.4	**8.6**
	Thunderstorm (days)	0	1	10	8	20	9	7	16	25	7	4	0	**107**
Mean	Cloudiness (0-10)	**2.6**	**2.2**	**2.2**	**3.7**	**6.4**	**7.7**	**8.1**	**8.0**	**7.3**	**6.13**	**4.71**	**2.7**	4.9
	Visibility (km.)	**6.5**	**5.0**	**5.2**	**6.6**	**11.0**	**12.7**	**12.4**	**12.4**	**11.3**	**10.0**	**9.6**	**7.9**	9.2
	Thunderstorm (days)	**0.1**	**1.1**	**3.3**	**9.3**	**15.1**	**7.8**	**7.3**	**10.3**	**15.1**	**7.1**	**2.1**	**0**	-

Source: METEOROLOGICAL DEPARTMENT, 1992

Remarks: '-' means no data

Tab. 17: Cloudiness, visibility, and thunderstorm data of the study area II (in the Central Plain) between 1981 and 1990, measured at the Station 412001 (Location : 14° 35' Lat., 100°27' Long., 8.0 m above MSL).

Year		Jan.	Feb.	Mar.	Apr.	May	Jun.	Jul.	Aug.	Sep.	Oct.	Nov.	Dec.	**Annual**
1981	Mean Cloudiness (0-10)	-	-	-	-	-	-	-	-	-	-	-	-	**-**
	Mean Visibility (km.)	-	-	-	-	-	-	-	-	-	-	-	-	**-**
	Thunderstorm (days)	0	3	3	8	15	4	8	5	3	10	5	0	**64**
1982	Cloudiness (0-10)	-	-	-	-	-	-	-	-	-	-	-	-	**-**
	Visibility (km.)	-	-	-	-	-	-	-	-	-	-	-	-	**-**
	Thunderstorm (days)	0	1	3	6	10	11	6	3	12	10	4	1	**67**
1983	Cloudiness (0-10)	-	-	-	-	-	-	-	-	-	-	-	-	**-**
	Visibility (km.)	-	-	-	-	-	-	-	-	-	-	-	-	**-**
	Thunderstorm (days)	0	0	0	0	15	9	10	15	16	14	1	0	**80**
1984	Cloudiness (0-10)	3.5	4.8	3.7	5.9	6.3	7.9	7.4	8.4	7.6	6.0	3.4	2.7	**5.6**
	Visibility (km.)	8.2	7.7	7.3	8.1	9.1	10.3	10.6	9.9	10.3	10.3	9.1	10.0	**9.2**
	Thunderstorm (days)	0	6	3	7	12	8	13	11	12	7	1	0	**80**
1985	Cloudiness (0-10)	2.4	3.2	3.4	4.8	6.7	8.3	7.8	8.3	7.8	5.9	3.8	0.7	**5.3**
	Visibility (km.)	9.1	8.2	7.8	8.2	10.7	11.0	10.9	11.2	10.5	10.9	11.4	10.4	**10.0**
	Thunderstorm (days)	3	1	0	6	11	3	8	5	13	11	2	0	**63**
1986	Cloudiness (0-10)	1.5	2.5	1.9	3.9	6.1	6.4	7.6	7.4	6.6	7.1	3.9	2.8	**4.8**
	Visibility (km.)	7.8	7.4	7.1	7.4	9.4	11.0	10.8	11.3	10.7	10.2	10.6	9.3	**9.4**
	Thunderstorm (days)	0	0	0	13	8	8	5	8	10	9	0	0	**61**
1987	Cloudiness (0-10)	2.2	2.5	3.4	4.4	5.4	7.4	7.4	7.4	7.6	5.4	6.1	2.3	**5.1**
	Visibility (km.)	7.9	6.0	6.6	6.5	8.9	9.7	9.3	10.0	9.7	9.9	11.1	8.5	**8.7**
	Thunderstorm (days)	0	0	1	9	12	12	8	7	14	12	4	0	**79**
1988	Cloudiness (0-10)	-	-	-	-	-	-	-	-	-	-	-	-	**-**
	Visibility (km.)	-	-	-	-	-	-	-	-	-	-	-	-	**-**
	Thunderstorm (days)	-	-	-	-	-	-	-	-	-	-	-	-	**-**
1989	Cloudiness (0-10)	3.5	3.0	4.0	3.1	6.6	7.8	7.9	7.9	7.6	5.4	3.1	1.0	**5.1**
	Visibility (km.)	7.7	6.2	6.8	7.1	9.0	9.9	9.6	9.9	9.3	9.7	9.7	7.2	**8.5**
	Thunderstorm (days)	1	0	3	3	12	4	7	8	10	5	2	0	**55**
1990	Cloudiness (0-10)	2.5	2.9	3.4	4.5	7.3	8.2	8.2	7.9	8.3	6.6	4.1	1.4	**5.4**
	Visibility (km.)	6.5	5.8	6.2	6.9	9.1	9.9	9.8	10.0	9.3	9.0	9.5	9.4	**8.5**
	Thunderstorm (days)	0	0	4	4	15	6	4	8	14	9	4	0	**68**
Mean	Cloudiness (0-10)	**2.6**	**3.2**	**3.3**	**4.4**	**6.4**	**7.7**	**7.7**	**7.8**	**7.6**	**6.1**	**4.1**	**1.8**	**5.2**
	Visibility (km.)	**7.9**	**6.9**	**7.0**	**7.4**	**9.4**	**10.3**	**10.2**	**10.4**	**10.0**	**10.0**	**10.2**	**9.1**	**9.1**
	Thunderstorm (days)	**0.4**	**1.2**	**1.9**	**6.2**	**12.2**	**7.2**	**7.7**	**7.8**	**11.6**	**9.7**	**2.6**	**0.1**	**-**

Source: METEOROLOGICAL DEPARTMENT, 1992

Remarks: '-' means no data

Tab. 18: Cloudiness, visibility, and thunderstorm data of the study area III (in the Korat plateau) between 1981 and 1990, measured at the Station 381003 (Location : 16° 29' Lat., 102°07' Long., 170.0 m above MSL).

Year		Jan.	Feb.	Mar.	Apr.	May	Jun.	Jul.	Aug.	Sep.	Oct.	Nov.	Dec.	**Annual**
1981	Mean Cloudiness (0-10)	-	-	-	-	-	-	-	-	-	-	-	-	**-**
	Mean Visibility (km.)	-	-	-	-	-	-	-	-	-	-	-	-	**-**
	Thunderstorm (days)	0	2	5	15	22	14	18	20	15	5	0	0	**116**
1982	Cloudiness (0-10)	1.1	2.7	2.6	4.9	5.7	6.7	7.1	8.6	7.9	4.6	3.1	1.9	**4.7**
	Visibility (km.)	6.5	7.0	6.1	7.3	8.7	10.9	11.4	11.2	9.6	10.9	10.5	8.2	**9.0**
	Thunderstorm (days)	0	4	7	6	12	7	10	5	10	13	1	0	**75**
1983	Cloudiness (0-10)	2.7	1.8	1.5	2.9	6.3	6.5	6.1	8.1	6.8	6.5	3.4	2.4	**4.6**
	Visibility (km.)	7.7	6.2	5.9	5.0	8.3	11.6	12.6	10.5	11.2	10.2	9.1	7.7	**8.8**
	Thunderstorm (days)	1	1	0	8	11	9	10	15	19	4	0	0	**78**
1984	Cloudiness (0-10)	2.2	2.5	2.4	4.8	6.4	7.6	7.2	8.0	6.4	5.0	3.7	1.8	**4.8**
	Visibility (km.)	7.7	7.7	8.1	8.5	11.1	11.7	11.4	11.4	11.6	10.6	8.6	9.1	**9.8**
	Thunderstorm (days)	0	2	4	13	17	15	16	12	18	6	1	0	**104**
1985	Cloudiness (0-10)	1.8	3.3	2.4	4.4	6.2	8.1	7.0	8.8	7.1	5.0	4.0	1.7	**5.0**
	Visibility (km.)	8.0	6.9	7.7	8.5	10.4	10.6	11.5	11.0	10.7	10.5	10.7	9.8	**9.7**
	Thunderstorm (days)	0	2	2	12	13	5	8	6	11	4	0	0	**63**
1986	Cloudiness (0-10)	1.6	1.5	2.0	4.7	6.8	6.8	7.5	7.5	5.0	5.0	3.4	2.5	**4.5**
	Visibility (km.)	9.5	6.0	5.9	8.8	10.5	11.4	11.5	11.5	10.7	9.7	9.3	8.5	**9.4**
	Thunderstorm (days)	0	0	0	12	13	12	4	13	10	4	0	0	**68**
1987	Cloudiness (0-10)	1.8	2.6	3.7	4.0	5.2	6.9	8.0	7.1	6.8	3.9	5.3	1.7	**4.8**
	Visibility (km.)	7.2	5.6	7.3	7.6	10.6	11.3	11.3	11.3	10.4	10.7	11.1	8.3	**9.4**
	Thunderstorm (days)	0	1	5	11	13	8	5	15	7	10	4	-	**-**
1988	Cloudiness (0-10)	-	-	-	-	-	-	-	-	-	-	-	-	**-**
	Visibility (km.)	-	-	-	-	-	-	-	-	-	-	-	-	**-**
	Thunderstorm (days)	-	-	-	-	-	-	-	-	-	-	-	-	**-**
1989	Cloudiness (0-10)	2.4	2.0	4.3	3.3	5.5	6.6	7.5	7.7	6.6	4.8	2.9	1.3	**4.6**
	Visibility (km.)	7.1	6.0	6.3	7.1	9.3	10.6	11.0	11.0	10.9	9.8	9.3	6.9	**8.8**
	Thunderstorm (days)	0	0	6	8	12	9	8	12	11	4	0	0	**70**
1990	Cloudiness (0-10)	2.2	2.7	3.6	3.7	6.5	7.9	7.8	7.3	6.8	6.2	3.8	1.4	**5.0**
	Visibility (km.)	6.8	6.7	7.7	7.9	8.1	9.5	10.0	10.9	10.8	9.6	9.5	7.8	**8.8**
	Thunderstorm (days)	0	7	6	6	17	12	13	16	14	3	2	0	**96**
Mean	Cloudiness (0-10)	**2.0**	**2.4**	**2.8**	**4.1**	**6.1**	**7.1**	**7.3**	**7.9**	**6.7**	**5.1**	**3.7**	**1.8**	**4.8**
	Visibility (km.)	**7.6**	**6.5**	**6.9**	**7.6**	**9.6**	**11.0**	**11.3**	**11.1**	**10.1**	**10.2**	**9.8**	**8.3**	**9.2**
	Thunderstorm (days)	**0.1**	**2.1**	**3.9**	**10.1**	**12.1**	**10.1**	**10.2**	**12.7**	**12.8**	**5.9**	**0.9**	**0**	**-**

Source: METEOROLOGICAL DEPARTMENT, 1992

Remarks: '-' means no data

Tab. 19: **Cloudiness, Visibility, and Thunderstorm data of the study area IV (in the coastal zone) between 1981 and 1990, measured at the Station 480003 (Location : 12° 22' Lat., 102°21' Long., 3.0 m above MSL).**

Year		Jan.	Feb.	Mar.	Apr.	May	Jun.	Jul.	Aug.	Sep.	Oct.	Nov.	Dec.	**Annual**
1981	Mean Cloudiness (0-10)	-	-	-	-	-	-	-	-	-	-	-	-	-
	Mean Visibility (km.)	-	-	-	-	-	-	-	-	-	-	-	-	-
	Thunderstorm (days)	1	5	4	18	21	11	17	6	17	12	4	0	**116**
1982	Cloudiness (0-10)	2.3	4.9	6.3	6.3	7.8	9.2	9.0	9.5	9.1	7.0	6.0	4.5	**6.8**
	Visibility (km.)	5.8	5.3	5.6	6.9	7.8	7.2	7.0	6.6	6.8	7.7	8.9	7.5	**6.9**
	Thunderstorm (days)	0	3	7	14	16	10	9	6	10	15	8	0	**98**
1983	Cloudiness (0-10)	4.5	5.7	5.9	6.4	7.3	8.2	8.4	9.2	8.3	8.6	6.3	4.6	**7.0**
	Visibility (km.)	5.6	5.3	5.1	4.5	7.7	7.6	7.3	6.7	7.6	6.9	6.9	7.2	**6.5**
	Thunderstorm (days)	0	0	1	1	18	8	18	15	16	18	3	0	**98**
1984	Cloudiness (0-10)	5.4	6.7	6.3	7.7	8.4	9.1	8.7	9.1	8.6	7.1	5.1	5.2	**7.3**
	Visibility (km.)	6.0	5.6	6.3	7.7	8.2	7.3	8.0	6.3	7.4	7.4	7.6	7.7	**7.1**
	Thunderstorm (days)	0	3	7	18	19	5	10	7	16	8	5	1	**99**
1985	Cloudiness (0-10)	4.6	6.5	6.9	7.8	8.4	9.3	8.7	8.9	9.0	8.2	7.2	4.2	**7.5**
	Visibility (km.)	6.8	6.9	6.5	6.9	8.0	6.8	7.2	6.7	7.6	7.2	8.0	7.9	**7.2**
	Thunderstorm (days)	2	3	8	15	19	5	13	9	10	15	4	1	**104**
1986	Cloudiness (0-10)	5.1	6.2	6.7	7.4	8.4	8.5	8.9	9.4	8.8	8.2	6.8	4.6	**7.4**
	Visibility (km.)	6.7	6.5	6.4	6.3	6.6	7.1	6.9	6.0	6.9	7.1	7.4	7.9	**6.8**
	Thunderstorm (days)	0	0	0	11	9	9	7	12	10	10	3	0	**71**
1987	Cloudiness (0-10)	4.9	5.9	6.7	7.1	7.8	8.6	8.6	8.6	8.6	7.3	7.8	3.8	**7.1**
	Visibility (km.)	7.2	6.2	6.4	6.1	7.3	6.8	6.2	7.0	6.6	7.0	7.9	6.6	**6.8**
	Thunderstorm (days)	0	1	2	9	15	12	10	18	17	15	7	0	**106**
1988	Cloudiness (0-10)	-	-	-	-	-	-	-	-	-	-	-	-	-
	Visibility (km.)	-	-	-	-	-	-	-	-	-	-	-	-	-
	Thunderstorm (days)	-	-	-	-	-	-	-	-	-	-	-	-	-
1989	Cloudiness (0-10)	6.2	6.6	6.5	6.1	8.2	8.4	8.4	9.0	8.6	7.3	4.7	3.1	**6.9**
	Visibility (km.)	6.2	5.7	6.3	7.1	7.4	7.6	7.4	7.1	7.4	6.9	7.3	6.0	**6.9**
	Thunderstorm (days)	0	0	4	6	14	10	11	9	11	11	2	0	**78**
1990	Cloudiness (0-10)	4.5	5.7	6.1	6.1	7.9	9.0	8.4	9.0	8.7	7.6	5.3	3.5	**6.8**
	Visibility (km.)	5.6	6.0	6.3	6.6	6.7	6.3	6.7	6.1	6.6	6.6	7.0	6.5	**6.4**
	Thunderstorm (days)	1	0	6	6	15	7	8	14	15	10	2	0	**84**
Mean	Cloudiness (0-10)	**4.7**	**6.0**	**6.4**	**6.9**	**8.0**	**8.9**	**8.6**	**9.1**	**8.7**	**7.7**	**6.2**	**4.2**	**7.1**
	Visibility (km.)	**6.2**	**5.9**	**6.1**	**6.5**	**7.5**	**7.3**	**7.1**	**6.6**	**7.1**	**7.1**	**7.6**	**7.2**	**6.9**
	Thunderstorm (days)	**0.4**	**1.7**	**3.6**	**10.9**	**16.2**	**8.6**	**11.4**	**10.7**	**13.6**	**12.3**	**4.2**	**0.2**	**-**

Source: METEOROLOGICAL DEPARTMENT, 1992

Remarks: '-' means no data

Using the data of Tab. 16 to 19, the cloudiness, visibility, and thunderstorm of each study area were analyzed and plotted into the form of graphic relationship between their values and time (month). The results are shown as the following figures.

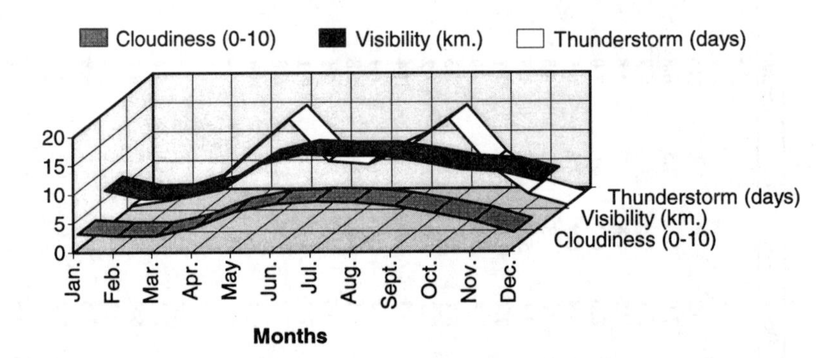

Fig. 24: **Average monthly cloudiness, visibility, and thunderstorm data of the study area I (Source: Prepared by the author using the data of the Meteorological Department, Bangkok).**

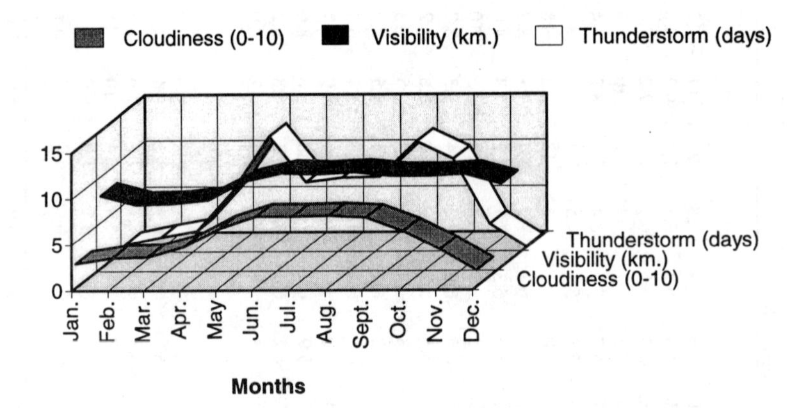

Fig. 25: **Average monthly cloudiness, visibility, and thunderstorm data of the study area II (Source: Prepared by the author using the data of the Meteorological Department, Bangkok).**

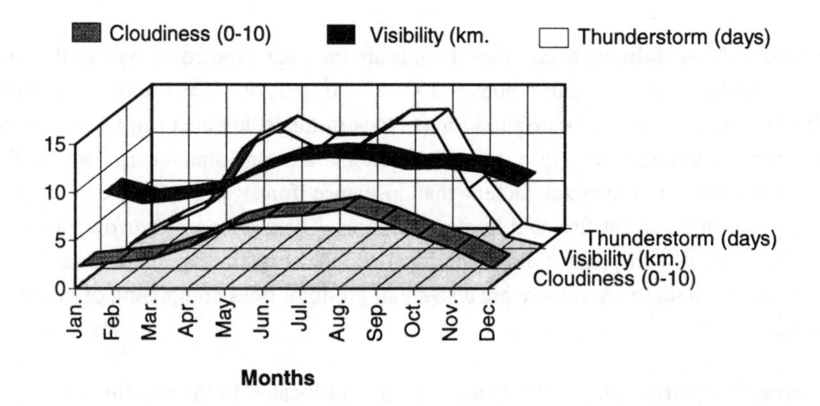

Fig. 26: Average monthly cloudiness, visibility, and thunderstorm data of the study area III (Source: Prepared by the author using the data of the Meteorological Department, Bangkok).

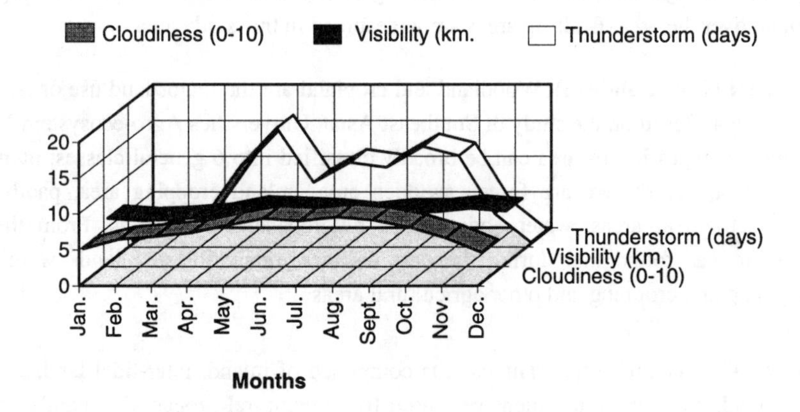

Fig. 27: Average monthly cloudiness, visibility, and thunderstorm data of the study area IV (Source: Prepared by the author using the data of the Meteorological Department, Bangkok).

From data analysis, it can be concluded that the low cloudiness period occurs from November to April in all study areas. Cloud-free satellite data could be obtained in this period. Thunderstorm curves show a bimodal distribution with the first peak around June or July and the second peak in October to November. Field work between these periods might be inconvenient. On the other hand, these periods show a good visibility.

4.6 Original land use and land cover types

Study area I (Mountainous area). Forest lands are the main land cover type of this area. They cover the higher areas (usually above 300 m). Based on forest type classification by SMITINAND et al. (1978), the original forest types in this study area can be mainly classified into 3 types, namely dry evergreen, mixed deciduous, and dry dipterocarp forests. There are, however, several environmental factors that influence forest vegetation; e.g. slope, aspect, elevation, soil types, forest fire, and human activities. There are some narrow basin areas that are mostly covered with paddy fields. The local people normally use these areas to grow rice and other crops. In these areas there are also small group of trees comprising of grasses, shrubs, and bushes.

Study area II (Central plain). This area is plain and located in the middle of the monsoonal zone of this region. In addition, there are rivers and many irrigation canals used for agricultural purposes in this area. Because of its fertile soil and its abundance in water resources, the people in this area earn their living by mainly growing rice. Field crops such as sugarcane were also found in particular where the areas are relatively higher and not flooded all the year. Furthermore, home gardens, cultivated with mangos, coconuts, bananas, and other vegetables can be found over the area. Built-up areas are widespread in this study area.

Study area III (Korat plateau). Woodland and cropland are the major land use or land cover types in this area. Based on the study of Southeast Asian Universities Agro-ecosystem Network (1987), land use types in this area can be broadly classified into 6 general classes; namely dry evergreen forest, dry dipterocarp forest, swidden area, upland cropping area, paddy fields, villages or settlement areas, water bodies, rock outcrops and bare lands. From the study referred to, it was found that during the past decades, many forested lands were mostly converted to upland cropping and other agricultural areas.

Study area IV (Coastal zone). This area is comprised of inland, inter-tidal land, and water surfaces. Inland, most areas are intensively used for agricultural production, mainly orchards, rubber plantations, and others crops. The inter-tidal zone, along the shoreline and the littoral plain were covered with mangrove forest. Mangrove forests in this area generally occupy the sheltered muddy and inland shore. Based on the study of SILAPATHONG (1992), it was found that the total mangrove forest area was seriously reduced and mostly converted into shrimp ponds.

5 METHODOLOGY AND TECHNIQUES

As stated in the literature review, information on land use and land cover is required in many studies concerned with natural resources planning, land use planning, and policy development. It is commonly a prerequisite for monitoring and modeling land use, land cover and environmental change. To enhance the availability of land use and land cover data sets, this study incorporates satellite image processing, terrain surface modeling, and geographic information system in order to prove the benefits of land use and land cover studies in the selected sites in Thailand. The physical characteristics of the study areas, described in the previous section, were primarily reviews at the beginning phase of the study. This is because these factors normally have a relationship with land covers on land. Climatic data were taken into account because of the interaction between land use types and the climate. Figure 28 shows a conceptual overview of analytical procedures and strategies applied in this study. In order to fulfill the purposes of the study, the following approaches were performed.

5.1 General descriptions of proposed procedures for land use and land cover classification by means of satellite digital image processing

In fact, the characteristics of satellite data and circumstances for each study can vary greatly. The review of previous studies particularly relevant to the application of satellite data analysis for land use and/or land cover classification were intensively done. However, the methodology of previous studies could not been directly adopted to use in this study without modifying. The success of previous studies were considerably used as guidelines for doing this research. Some interesting techniques were primarily tested before being adopted. To deal with the computer-aided classification systems, the understanding of the alternative strategies for satellite digital image processing is significantly necessary. Furthermore, the principle of Remote Sensing was always kept in mind during perform the procedures of the study. The important principles as well as significant theorem were referenced and applied throughout the study.

The first step of the proposed procedure was to design an appropriate land use and land cover classification system (scheme). To do this, existing land use and land cover classification systems referred to the literature were evaluated and identified in order to formulate an optimal system for a classification. Using the existing systems as guidelines, an initial development of a new system was conducted. During the developing stage, an evaluation of this system were based on the suitability of using the system with the high resolution satellite data such as Landsat TM data. It was designed to discriminate better significant features of various land use and land cover categories, representative of each study area. Besides, the capability of satellite image processing techniques in capturing land use and land cover data of selected test sites were performed and proven. In the final stage of the methodology, the classification system was optimized again, based on the success of image classification.

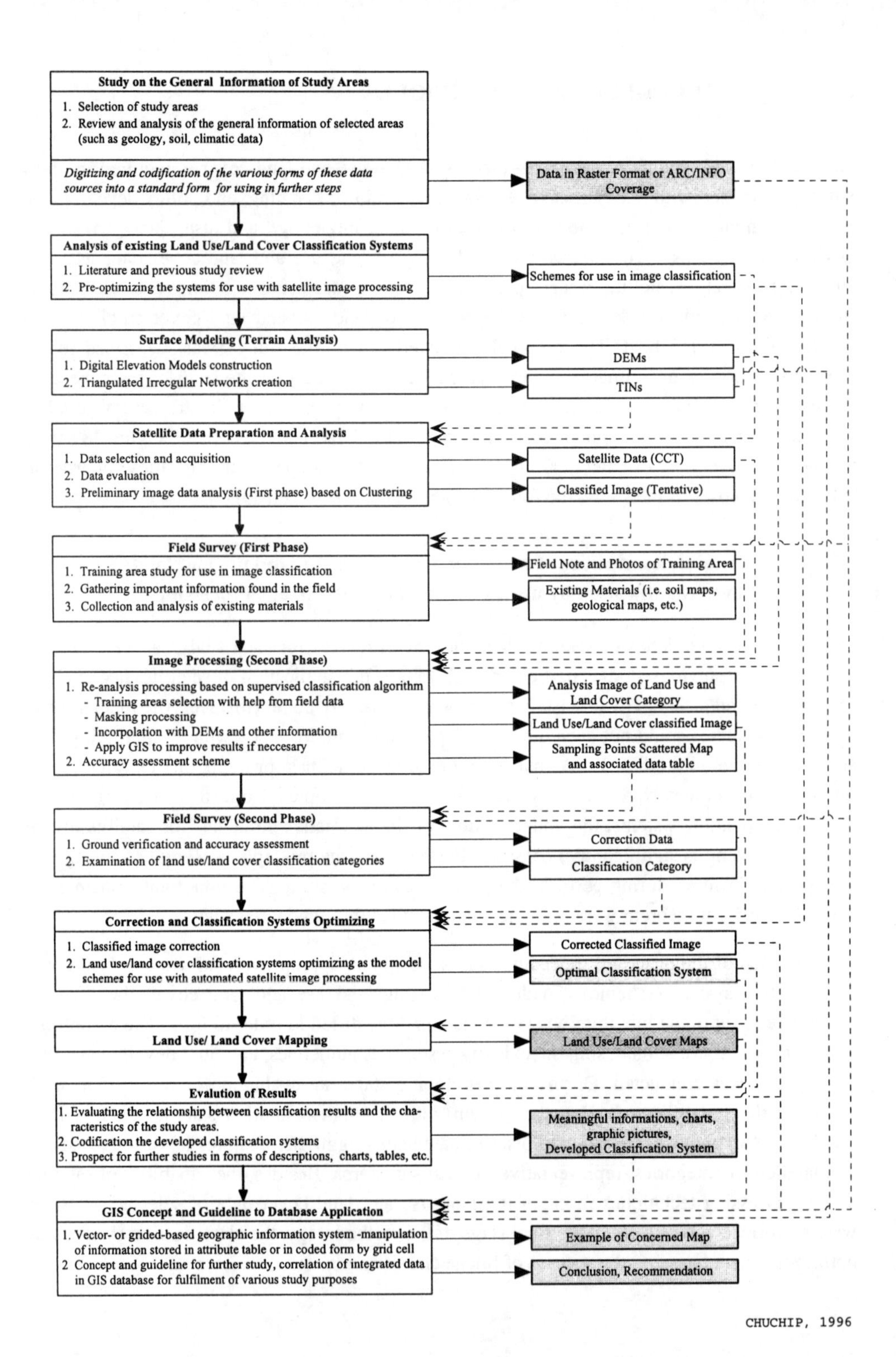

Fig. 28: **Conceptual overview of analytical procedures and strategies employed as well as the data flow applied in the study (Source: Author).**

5.1.1 INITIAL LAND USE AND LAND COVER CATEGORIES

Land use and land cover classification are commonly necessary as basic information for formulating many important planning activities concerned with the surface of the earth. This means that most of land use and land cover types occurring on land are actually important for many studies. Thus, categorization of land use and land cover should cover and be based on existing land use and land cover types, and on the natural conditions of the area. A classification system performed in the first stage of this study included possible categories subject to existing status of land cover types as well as active land use types. However, it is necessary to consider the possibility of identifying these objects from satellite data. Theoretically, earth surface features of interest can be identified effectively on the basis of their spectral characteristics. It has, however, to be kept in mind that some features of interest cannot be spectrally separated. Thus, an evaluation of satellite digital data must be done for a better understanding of the spectral characteristics of the particular features, together with understanding the definition of classification categories.

Every object on the earth's surface reflects electromagnetic rays of various wavelengths in its own characteristic way. This characteristic reflectance allows recognition and separation of objects. Figure 29 shows typical spectral reflectance curves for three types of earth features; namely healthy green vegetation, dry bare soil, and clear water.

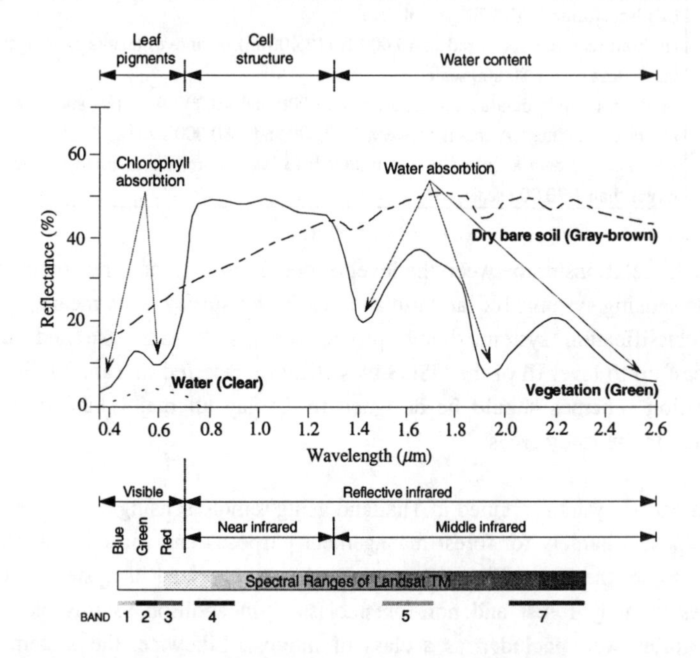

Fig. 29: Typical spectral reflectance curves for vegetation, soil, and water in relationship with spectral bands of Landsat Thematic Mapper (Adapted from LILLESAND & KIEFER, 1994; and SWAIN & DAVIS 1978).

Considering the curves in Fig. 29, it is obvious that satellite data from Landsat TM can give very useful results for land cover classification. Noticeably, absorbtion property of water is very distinctively characteristic. Locating or delineating water bodies with satellite data can be done easily in near-infrared wavelengths of the Landsat sensors. Experience has also shown that the spectral classification of the digital satellite imagery can achieve a satisfying result with nearly 100 percent accuracy for the discrimination of water and land. Thus, initial land use and land cover categories were correlated to this notice.

Based on existing literature, it seems that many existing land use and land cover systems were designed or developed for visual interpretation of remotely sensed data obtained at various scales and resolution by sensors. The relationship is illustrated in the four levels of the US Geological Survey Land use and Land Cover Classification System as shown in Tab. 1. It is important to keep in mind that there is a relationship between the level of detail in a classification system and the spatial resolution of remote sensor systems used to provide information.

Tab. 20 The four levels of the US Geological Survey land use and land cover classification system and the type of remotely sensed data typically to provide the information (Source: JENSEN, 1986).

Classification level	Typical data characteristics
I	Landsat (formerly ERTS) type of data
II	High-altitude data acquired at 40,000 ft (12,400 m) or above; results in imagery that is less than 1:80,000 scale
III	Medium-altitude data acquired between 10,000 and 40,000 ft (3,100 and 12,400 m); results in imagery that is between 1:20,000 and 1:80,000 scale
IV	Low-altitude data acquired below 10,000 ft (3,100 m); results in imagery that is larger than 1:20,000 scale

Figure 30 shows the relationship between the level of detail required and the spatial resolution of various remote sensing systems for land use and land cover survey. This means, in our case, that a desired classification system should provide details of land use and land cover information at least up to level III of the USGS classification referred in Tab. 20. Furthermore, desired classification schemes should be designed to display all major land uses and land covers encountered in the study areas.

Normally, classification systems applied in Thailand using remote sensing data were developed for two main purposes: namely for forest management purposes and for agricultural land use purposes. For example, the Royal Forest Department has categorized its system starting from two main classes, namely forest and non-forest classes. In addition to this categorization, agricultural categories were included as a class of interest. Likewise, the systems of other departments concerned with agricultural purposes were mainly oriented to agricultural categories while a forest class was broadly included as a main category into those systems. It can be assumed that an optimal classification system must be included both agricultural and

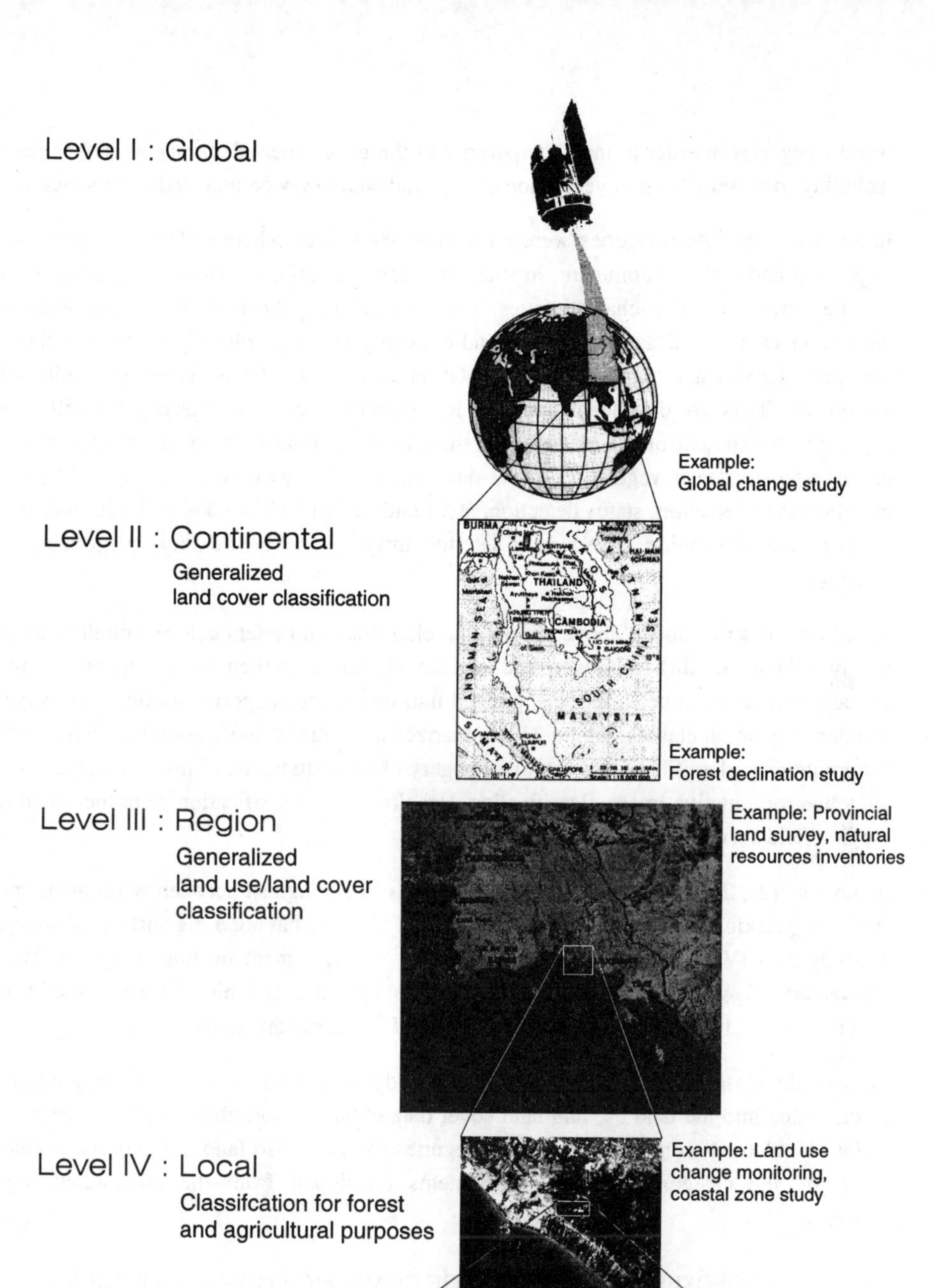

Fig. 30: **Relationship between the level of detail required and the spatial of various remote sensing systems for land use classification (Adapted from JENSEN, 1986).**

forest categories in order to meet the purpose of the government departments. Other categories including some semi-natural vegetation classes and water may be included as subclasses.

In any case, satellite imageries were taken over land areas where reflecting objects, such as vegetation and soil, are dominant. In such case, the spectral reflectance of vegetation and soil are the most distinctive characteristics. Thus, considering the ease of the discrimination of objects on such digital data, vegetation indices can play a big role. Figure 29 also shows that the spectral reflectance properties of vegetation canopy vary from visible to middle infrared wavebands. They are dependent upon the leaf pigments (e.g. chlorophyll), the cell structure, and the water content of the leaves. That means that optimal wavelength sensors of satellites can be used to detect vegetation in various manners. For example: for species distinctions, growing stage detection, status detection, etc. Landsat TM can be adopted to do that. It can be assumed that vegetation covers in a satellite imagery can be flexibly categorized for our purposes.

Based on notes mentioned above, an optimal classification system can be initially categorized into two classes starting from water and land classes, based on their spectral features. And then, the cover on land can be further categorized into two main categories starting from vegetation and non-vegetation classes. Figure 31 summarizes the initial classification hierarchy developed for preliminary classification of satellite imagery of each study area. This system was used as a guideline for tentative image classification. Results of the classification were then used during the first field survey.

In other words, the first step in pre-classification is to distinguish between water areas and land surfaces. Masking is proposed to be used with an imagery having a big surface of water, such as study area IV in the coastal zone. The result of experiment on this study area was very satisfactory. However, the water surface detection algorithm is limited in the case of very thin water coverage, for example in case of study area I in the mountainous area.

In short, the classification systems of the four study areas were based on existing systems and incorporated into the land use and land cover data obtained from the satellite imagery classification. In addition, some classification categories used are also land use activity oriented and based on the categories of existing systems developed from the government agencies concerned.

5.1.2 PROPOSED SATELLITE IMAGE CLASSIFICATION STRATEGIES

Digital satellite image classification is the process of assigning image pixels to classes. In principle, each class derived is treated to have homogeneous pixels. In practice, however, some researchers may sometimes treat heterogeneous pixels into one needed class. Automated classification of digital image is generally the art of grouping pixels which have uniform spectral values. Generally, the analyst will try to treat or match these spectral classes to his specific categories of mapping. Although the classification scheme above (Fig.31) is composed of desired categories, the possibility to derive all of them from satellite image classification has to be investigated.

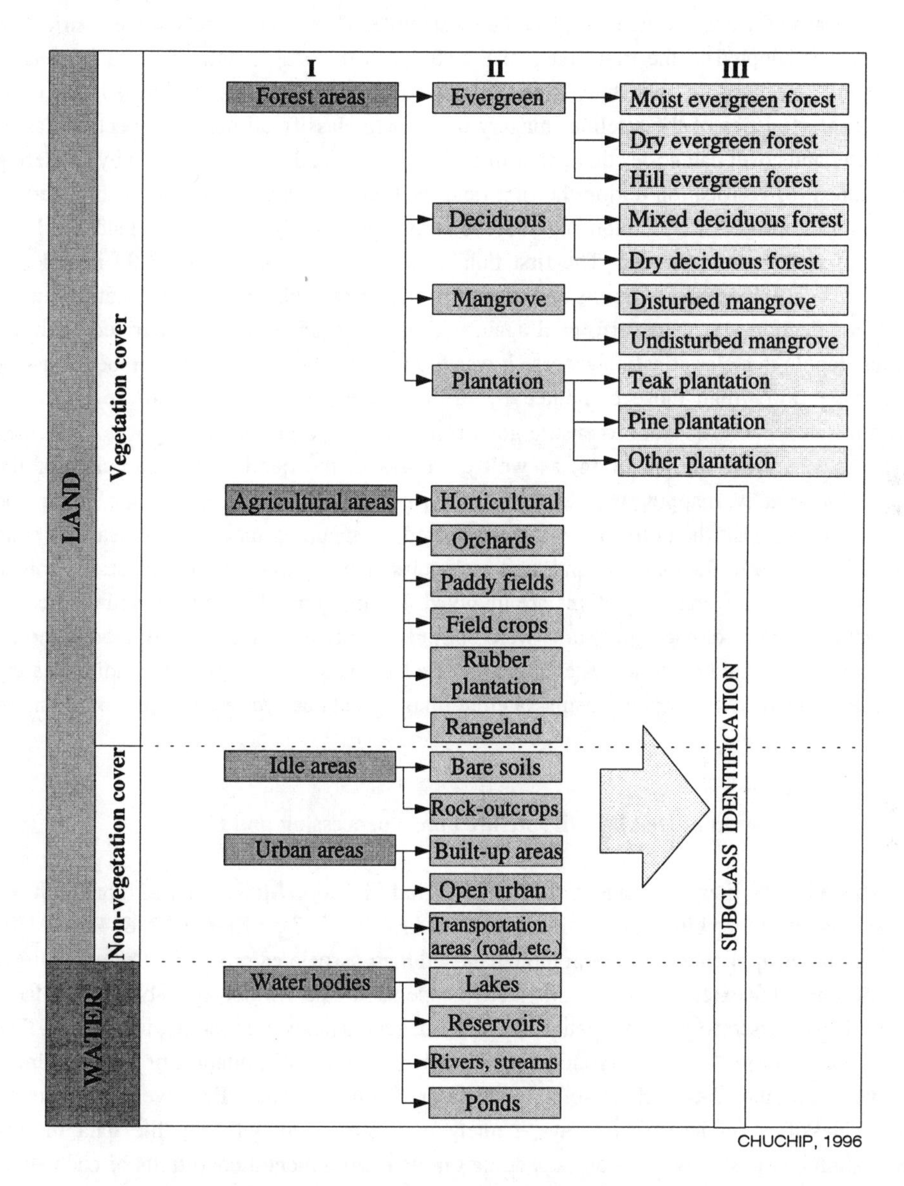

Fig. 31: The hierarchical land use/land cover types aggregated from existing classification systems with consideration of the possibility of digital satellite data analysis (Source: Author).

A digital image classification is sometimes a case of trial and error to achieve the best classification. At the moment the unsupervised classification is the most frequently used technique for automatic recognition of the objects on the ground, in particular concerning land use applications. The main aim is to detect and recognize objects by their spectral features taken by means of sensors which are placed on a satellite. Thus, an unsupervised classification algorithm was adopted in the first phase of the study. In this stage, statistics of all pixels of satellite imagery used in the study were calculated. **Clustering** was performed based on spectral characteristics of the satellite imagery in order to classify all possible spectral classes of land use/land cover categories that occur in each study area. All classes derived by clustering were mapped for verification during the first field visit. Occasionally, **Masking** technique was conducted to improve classification accuracy, in particular in study areas where land and large surfaces of water bodies existed. The first field visit was undertaken in 1993-94 in order to identify the study areas, and to define and determine the spectral classes of classifications in the field. Classification was then performed again with the help of information derived from the previous step. In this classification process it was found that in most cases the number of useful classes were fewer than those computed by the first classification in which the clustering algorithm was applied. In order to ensure good results a second ground survey was performed in 1994 to verify the classification, as well as to assess the quality and reliability of the classification. Finally, mapping was done using the final results of the classification. It must be noted at this stage that the number of classes selected for mapping may not necessarily be the same as those used in the land use and land cover classification systems optimization. This is due to the fact that different objectives are involved. In this study, the land use and land cover classification systems of each study area were optimized with all derived information based on the purpose of the study. These systems can be used as standards for further studies dealing with land use/land cover mapping using satellite imagery data derived particular from Landsat TM.

5.2 Integration of DEMs with satellite image processing and GIS

As mentioned in the literature review, the use of Digital Elevation Models (DEMs) has shown a good success in many kinds of studies, e.g. SCHARDT (1987), LEPRIEUR & DURAND. (1988) involving application of topographic data and their associated products. The own DEMs were developed for extensive use in this study, since such data are not available in Thailand. The DEMs were used to prepare spatial information, such as elevation, slope, and aspect. The DEMs were also used as a background for displaying thematic information or for combining relief data with other data such as soils, land use or vegetation. The DEMs were also used to improve classification in various ways, namely in combination with satellite data in pre-classification and post-classification depending on the environmental conditions of each study area.

Normally, the use of DEM prior to classification involves a division of the study image into smaller areas or strata based on intended criterion or rule, so that each stratum can be processed independently. The purpose of stratification is to increase the homogeneity of the data sets to be classified. In this study, this technique has not been performed because there were not enough information for specific suitable criterion or rule. Conceptually, the use of DEM during classification process can also be included in the classification algorithm by modifying the maximum likelihood decision rule. However, the image processing software used, ERDAS 7.5, has not afforded the ability to modify the decision rules or classification model. Thus, the use of DEM in the image processing process for this study is different. Integration of a DEM into image processing was tested in the case of study area I in the mountainous area. Shaded relief model was created from the DEM and used to correct shadow effects occurring in the satellite imagery. The methodology can be found in Section 5.4.6. DEMs were also used to create new image data, i.e. slope image, aspect image, shaded-relief image, and 3D terrain models. These DEM-product images were used as ancillary data for use in the training process of the supervised classification approach.

The use of DEM after the classification process in this study was based on Geographic Information System (GIS) concepts. The approach and techniques used here were performed to improve mapping accuracy. The problem occurring after classification were some miss-classified categories, such as mangrove stands included with natural forest on mountains, or water bodies occurring in high mountains. Shaded relief, slope and contour data were performed using the DEM. These data were then overlaid on the classified image with the overlay analysis in a grid-based geographic information system. Some miss-classified classes were considered to be grouped or merged into the appropriate existing classes.

In addition to those techniques, the display of three-dimensional (3D) perspective views was used as a visual simulation of the terrain in this study. Based on the position from which the view is calculated, it is possible to produce the graphic as if the viewer stands or hovers above the landscape. Perspective views are useful in graphic presentations, such as before and after views simulating the effect of proposed development. The generation of a 3D perspective view requires an elevation surface and a number of parameters that establishes how the view will be displayed (ERDAS, 1991a; ESRI, 1992b). These parameters include field of view, sun elevation, sun azimuth, the vertical exaggeration. In this study, 3D derived from DEM was displayed mostly in perspective view projection method instead of parallel view projection. Figure 32 shows a conceptual overview of using DEMs and theirs associated products in various ways employed in this study.

5.3 Construction of Digital Elevation Models and Triangulated Irregular Networks

Digital representations of topography and the variation of surface elevation over an area can be modelled in many ways, such as in form of grid matrices of elevation, series of parallel profiles, digitized contours, or triangulated irregular networks (TINs). In order to interpolate and display continuous surfaces needed in satellite image processing and GIS application in this study, the construction of both DEM and TIN was carried out. A Digital Elevation Model may be created by either digitizing contour lines from existing topographic maps, collecting

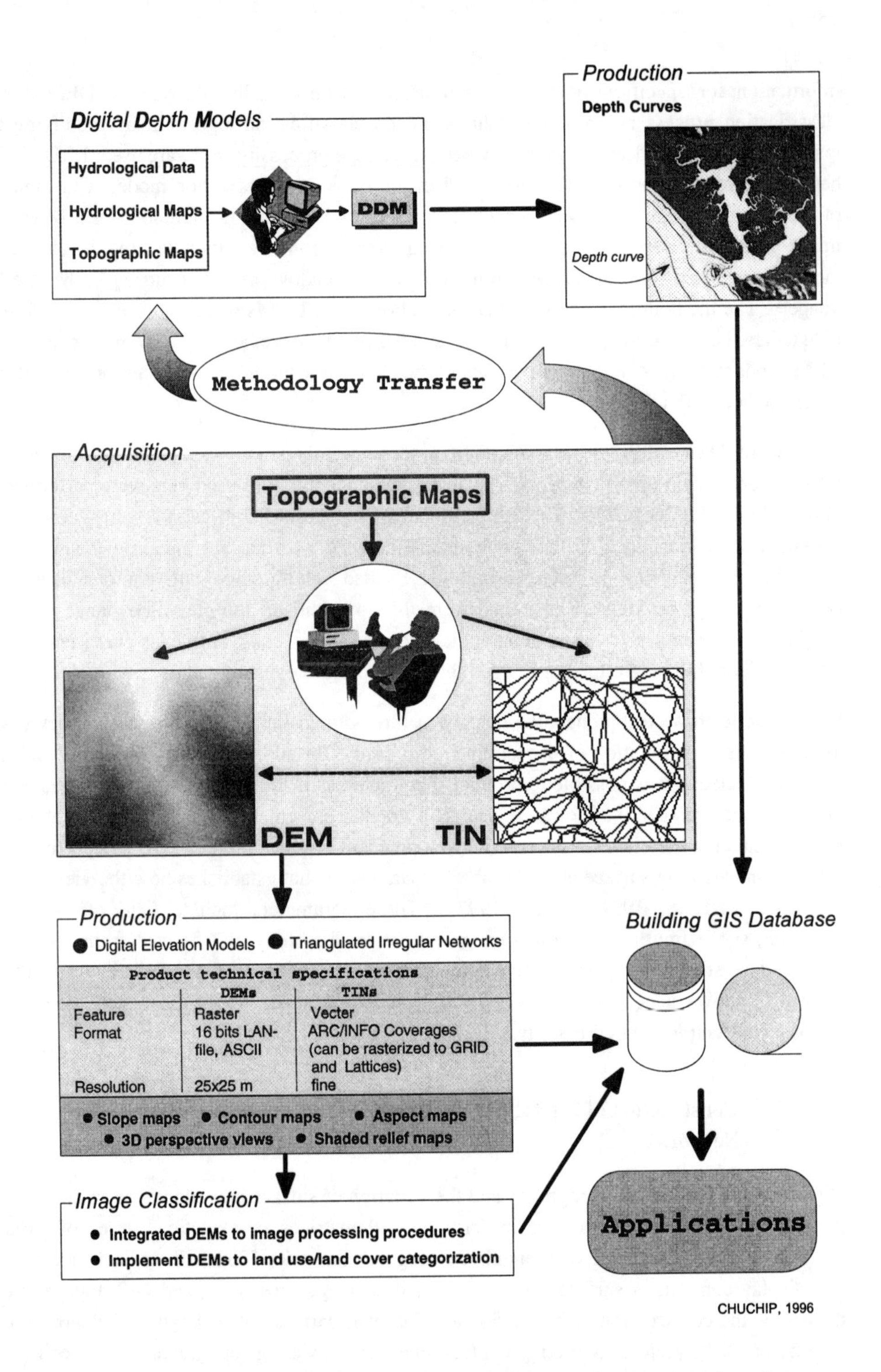

Fig. 32: **Conceptual overview of using DEMs and theirs associated products in various ways employed in this study (Source: Author).**

elevations by field surveys, or as a product of photogrammetric stereo-compilation (ERDAS, 1991a; ESRI, 1992b). A DEM is a common digital format containing an infinite number of values for representing a surface in a matrix of equally spaced sample points forming profiles across a surface. A TIN connects a set of irregularly spaced x-y-z locations and is composed of triangles, nodes and edges (ESRI, 1992a).

A DEM is suited for calculation of slope, aspect, sun intensity, shaded relief. It covers a regularly spaced, systematic sample of a surface. Thus, the DEM was adopted in this study in order to represent the real world land surface of the study areas and to provide topographic data in raster forms, e.g. contour lines, slopes and aspects. A TIN is, however, also useful for representing surfaces of some study areas that are highly variable, and contain discontinuities, breaklines, and no-altitude data areas. For the alternative use, the construction of both DEMs and TINs were implemented and carried out. DEMs were created by means of surfacing routine concepts in the ERDAS™ software, while TINs were constructed by taking advantage of the triangulated irregular network capabilities of the ARC/INFO® software.

5.3.1 **DEM** STRUCTURING

One of the most simple but efficient ways of constructing a DEM is the so-called interpolation method using elevation data from existing topographic maps (ERDAS, 1991). Interpolation is a process, whereby a value is derived for a new point based on the known values of points around it. Followings are the schemes of DEM structuring using the ERDAS version 7.5 software applied in this study.

(1) Getting topographic data

Elevation data can be derived by digitizing either manually or by automated raster scanning. In this study, contour lines have been digitized directly from topographic maps at a scale of 1: 50,000. Resulting data are the series of points called elevation or terrain data points. These elevation data are composed of a series of points with x, y, and z values. x and y values are grid coordinate values according to the coordinate system of topographic maps. z is the elevation, or altitude value at the point, where the coordinates x and y are located. Contour lines accepted in the digitizing process vary for each study area, depending on the topographic characteristics of the area and the restrictions of the software used (a DEM is usually stored in a 16-bit range of elevation values, meaning each pixel can have a possible elevation of -32,768 to 32,767). The digitized elevation data have to be sufficient but not redundant for use in the interpolation process. Generally speaking, the set of elevation points should be located and irregularly spaced as far part as necessary to accurately portray the variations of the terrain. Digitizing contours for surfacing processes is different from for the use in mapping. In digitizing contours or terrain data points for raster surfacing purposes, the goal is to space the points as evenly as possible over the entire map being digitized. Extra elevation points at the highest and lowest places on the map, along ridges, and along valleys and stream beds, were also included where they were found on the maps.

In this study, topographic data used for DEM structuring are derived from a digitizing operation of ARC/INFO. This is because these data will be also modified to use to create a TIN. Most of the digitizing routines in this study were performed on ARC/INFO. These captured data are called *coverage* that contain the arcs of digitized contour lines and some extra data points. ARC/INFO requires the construction of a topology to create spatial relationships between the features in a *coverage*. ARC/INFO assigns feature attribute tables as an information file associated with each feature type of its *coverage*. In topology constructing, the Arcs Attribute Table (AAT) and the Polygon Attribute Table (PAT) are created for a line *coverage* and a polygon *coverage*, respectively. For point *coverage*, the PAT is conceptually used as the attribute table in the same way as a polygon *coverage*. In this study, both lines and points were digitized and made into *coverages* by ARC/INFO. From digitized lines and points, a topology was separately constructed and performed to *coverages*. From the attribute table of the *coverage*, distinct GIS values were specified according to the elevation values. The GIS values are needed as z values in the surfacing routine of the ERDAS. The derived *coverage* must be converted to a digitized-polygons file format of ERDAS (a DIG file). From the experience in this study, this conversion often brought about unexpected or unforeseen problems. Converting an *arcs-coverage* into a DIG file can be easily success but conversion from a *points-coverage* to a DIG file can fail. Errors may occur with a *coverage* containing only point data, without any polygons. ERDAS assumes that a PAT-file form of the ARC/INFO *coverage* contains information in the first record of the attribute file on the external polygon or universe polygon. Subsequently, the ERDAS software will start reading and converting data from the second record of the attribute file, the PAT file of ARC/INFO. For a *coverage* containing only points without polygons, meaningful data are contained in the first record of the attribute file after its topology has been constructed (with BUILD command in ARC/INFO). As a result, the ERDAS software will ignore this first record, whenever a *point coverage* is converted to a DIG file. This will cause the coordinates and GIS values of each point be mismatched. Careless users may adopt this erroneous DIG file in their application although the error clearly occurs during the converting process. This error may be unnoticed, if the process of converting any ARC/INFO *coverages* to ERDAS's DIG-files is run automatically with a batch file. Such an error may cause unexpected results whenever the errorneous DIG file is utilized. However, such a problem is avoidable. In this case, a simple technique was applied to solve the problem and therefore make an attribute table file of *point coverage* usable. Each *point-coverage* of ARC/INFO being used in ERDAS was first edited by adding an extra record into its attribute table file. In this new record, zero values were assigned to all the relevant items. Then, items of the attribute table were sorted to rank the added zero-value items to be in the first record of the attribute file. As a result, this format file can be read error-free by ERDAS. The DIG files derived both from line and *point coverages* were merged for use in the next step.

(2) Generating a surface

In the ERDAS software, the digitized contours were stored as topographic data files. These files can be represented as a surface, or a DEM, in one band of an image data file, called LAN file. To do this, SORT and SURFACE programs in the Topographic Module of the software packages were used to generate an elevation surface. The SORT program was applied to create a point *block file* (with BLK extension) from a digitized contour file. The topographic data files are divided into a number of smaller processing blocks. The blocks are the square grid of an appropriate cell size calculated to be 1/10th of the length or width of the file. This grid is overlaid on the digital image of the digitized contours (see Fig. 33).

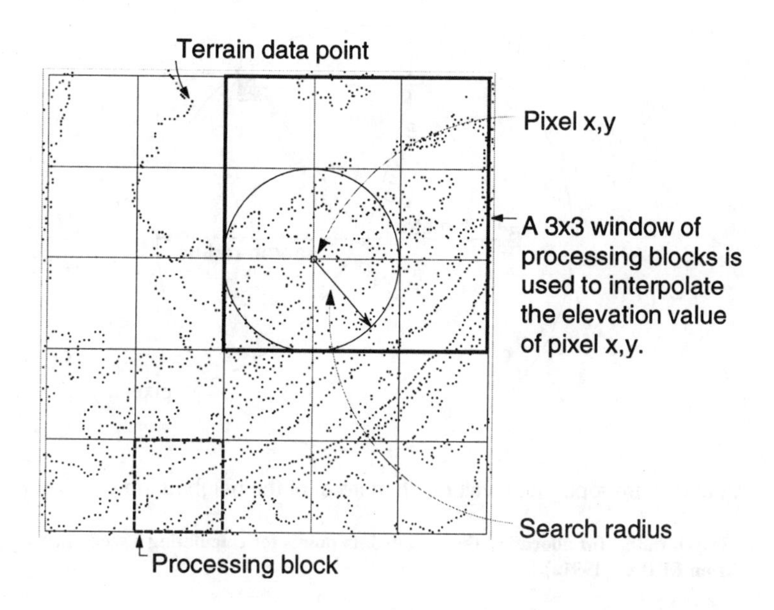

Fig. 33: Processing block for generating an elevation surface (adapted from ERDAS, 1991a).

The surfacing program uses only 170 terrain data points in each block; others will be ignored. Thus, while digitizing contours for the surfacing routine, it is not necessary to achieve a smooth contour line with dense data points. The SORT program initially calculates a block size of the grid as a square with each side being 1/10th of the length or width of the file, whichever is longer. This size is called a ***blocking factor***. The blocking factors used in the study were adjusted differently in each study area depending on the scattering of terrain data points.

To create a continuous surface stored in the form of a DEM file, the SURFACE command was conducted to process the terrain data points of the DIG file derived from the previous step. To minimize the amount of memory required during the SURFACE processing, a 3x3 window of the processing blocks was used to bound the terrain data point of the DIG file. The process operates only in the 3x3 window at any given time. The 3x3 window then moves through the entire file until a continuous surface has been calculated. To prevent a directional bias, a ***search radius*** is established. The search radius is used to specify the distance around each pixel in which the SURFACE program will search for terrain data points. The elevation value for the pixel is interpolated from the values of the terrain data points that fall within the search radius. If there are no elevation points found within the search radius, the value of the pixel is set to zero. The elevation values for any given pixels are extrapolated, based on the terrain data of surrounding points. The surface calculation for any given pixel uses a weighting function. The weighting function is based on the distance between the subject pixel and all other terrain data points, GIS or z value, in the input DIG file within a specified area, as illustrated in Fig. 34.

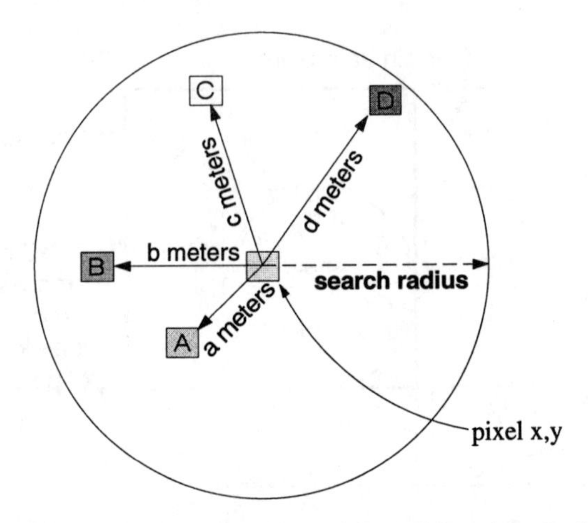

A, B, C, and D are topograhic values (in meters) of the neigbouring pixels of pixel x,y

Fig. 34: Search radius for allocating the terrain data points for calculating a seed pixel x, y (adapted from ERDAS, 1991a).

Weighting functions of the ERDAS topographic module for use in surfacing interpolation can be plotted as a graph shown in Fig. 35. The shape of the weighting function determines how the surface will look like.

These are 11 distance functions for calculation of weighting factors available in ERDAS revision 7.5:

1. $(1-Q)/Q$
2. $((1-Q)/Q)^2$
3. $\log(1/Q)$
4. $1-Q$
5. $(1-Q)^2$
6. $e(-0.5) \times (5Q)^2$
7. $e(-0.5) \times (10Q)^2$
8. $(1-Q^2)/Q^2$
9. $1-Q^2$
10. $((1-Q)/Q)^{0.1}$
11. $((1-Q)/Q)^{0.5}$

where

$Q = D/S$
D = calculated distance
S = search radius

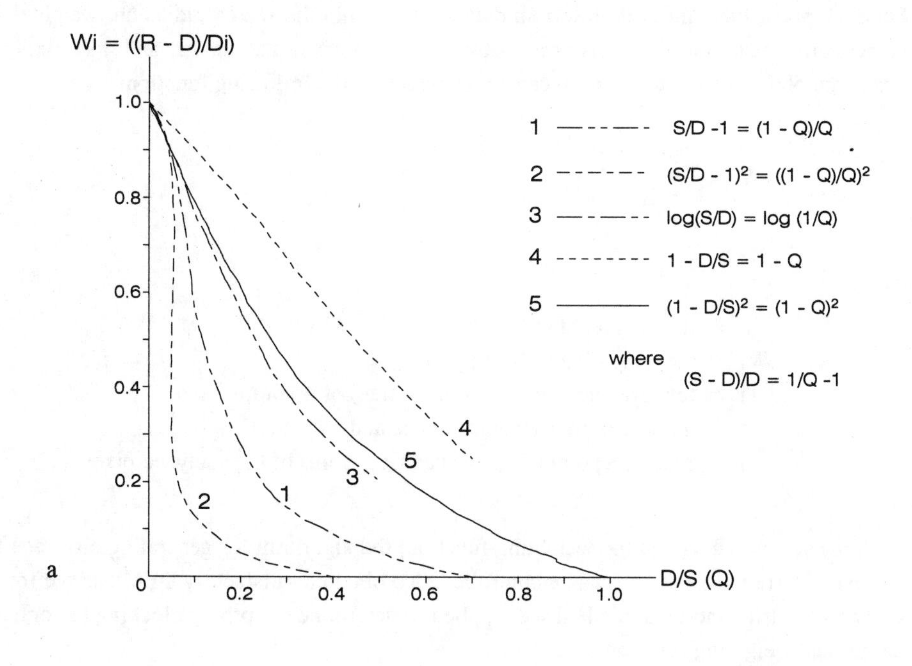

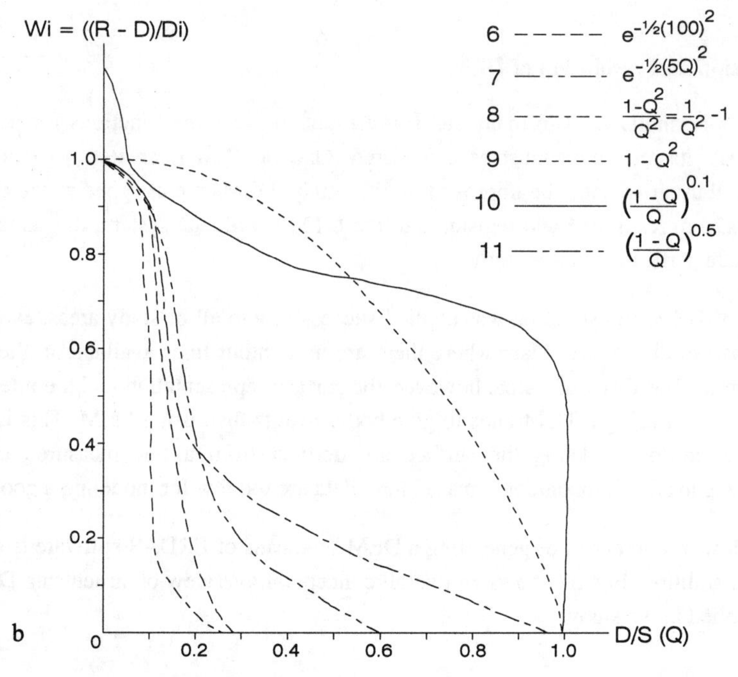

Fig. 35: Weighting options used for surfacing operation (ERDAS, 1991a).

For each pixel, the values of all terrain data points within the search radius are weighted by a value corresponding to the distance between each terrain data point and the pixel. The algorithm of the surface calculation can be expressed as the following function:

$$V = \frac{\Sigma W_i \times T_i}{\sum_{k=1}^{c} W_k}$$

Where

V = the output data file value,

W_i = the weighting factor of point i,

T_i = topographic value (GIS or z value) of point i,

c = number of topographic values, and

i represents a point within a specified radius of the analyzed pixel.

With the search radius and the weighting function, the algorithm for generating a surface from a series of terrain data points can be controlled. To obtain a satisfactory DEM surface for each study area, a trial-and-error method was applied to determine the proper blocking factor, search radius, and weighting function.

(3) **Registration and resolution of DEMs**

Since DEM is normally used in many studies relevant to modeling functions for performing spatial analysis, its resolution must be considered. Once a DEM is created, its resolution is established and usually cannot be improved. In this study, DEMs are assigned as the cell-based data with 25x25 m resolution and registered to the UTM coordinate system, the same as other geographic data performed in this study.

The method of DEMs construction was applied successfully to all of study areas, even in case of study area II in the Central Plain where there are no contour lines available in the relevant topographic map. For this study area, however, the surface representation and the interpolation of topographic data using a TIN trends to give better results than using DEM. This is because terrain data used for modeling the surface are derived from a few measured elevations contained in the available topographic map. These data are too few for modeling a good DEM.

An overall flow of operations of generating a DEM by means of ERDAS software is described in Fig. 36. In addition, Fig. 37 shows an overall conceptual overview of structuring DEMs and data flow applied in the study.

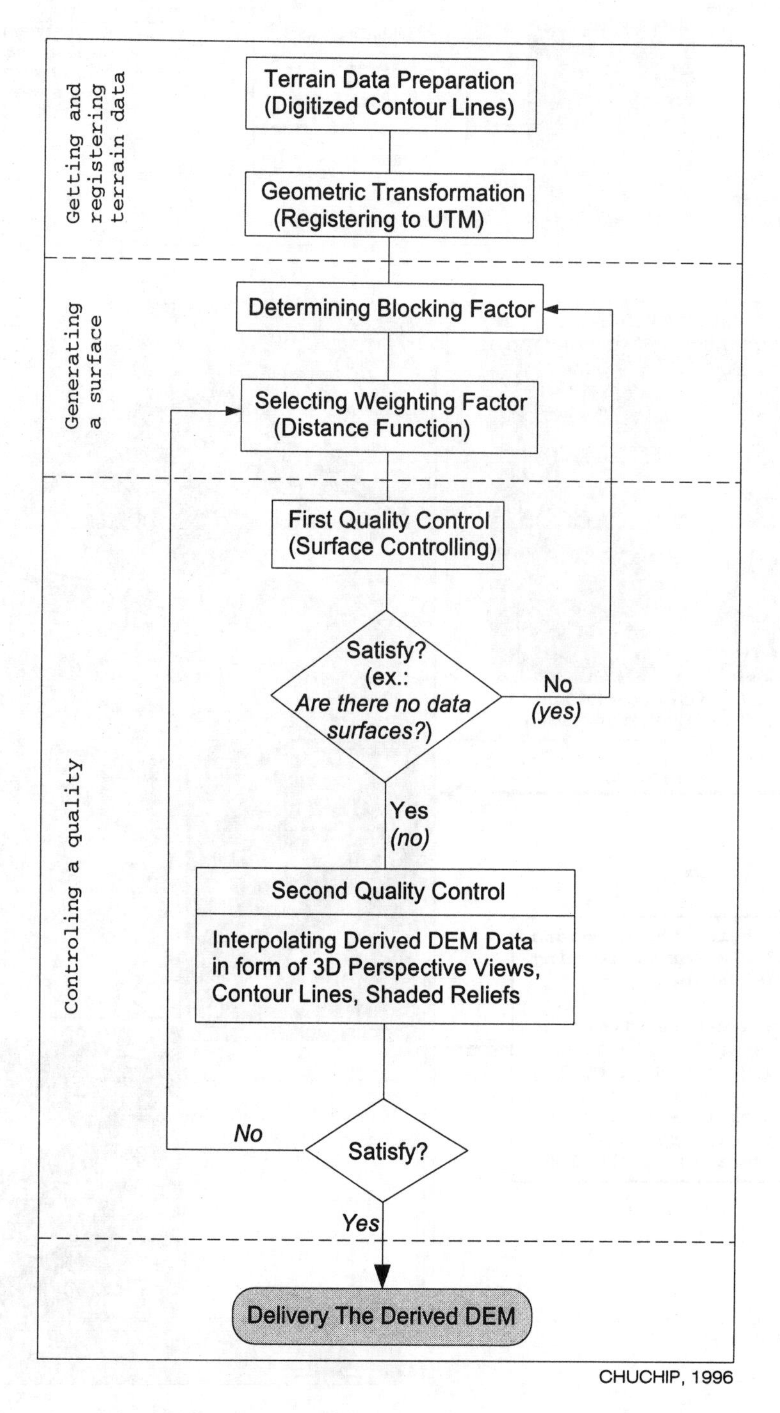

Fig. 36: **Flow of operations for DEM generating by means of ERDAS software concept applied in this study (Source: Author).**

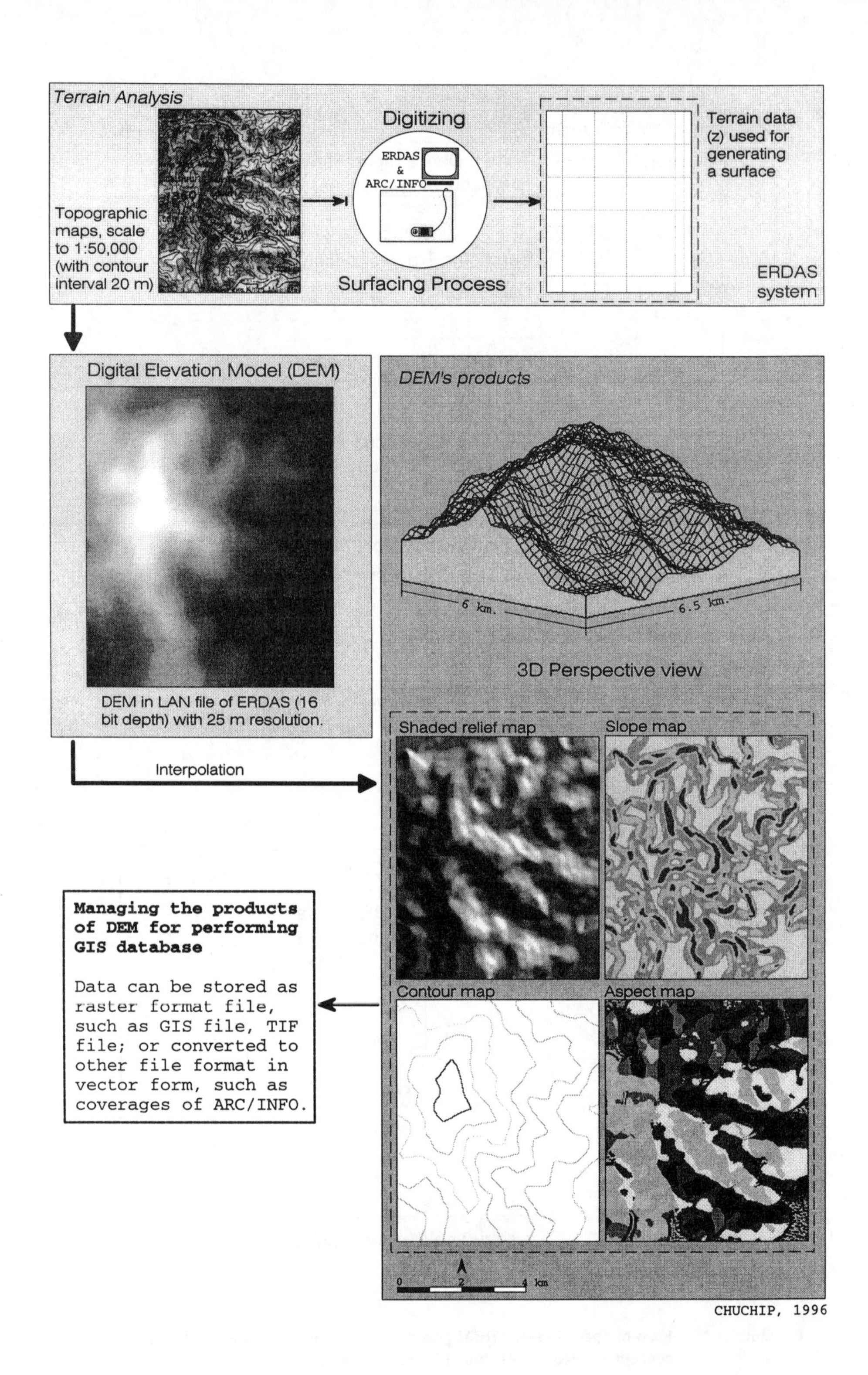

Fig. 37: **Conceptual overview of DEM construction and its products (Source: Author).**

5.3.2 TIN CREATION METHOD

The concept of TIN was also tested as an alternative means for representing surfaces efficiently in this study. Terrain modeling with TIN can use nodes of the digitized contours as irregularly spaced point samples to generate a sheet of continuous, connected triangular facets that are the main structure of a TIN. The data capturing process for a TIN can also specifically follow ridges, stream courses, and other important topological features that can be digitized to the accuracy required. To derive such data, contour lines together with elevation spots that represent the critical points along ridge-lines, the top of the mountains from topographic maps, have been digitized for each study area. In some cases, some more altitude data measured directly in the field with the help of a Global Positioning System (GPS) are also necessary. However, this measure was done only, where the reference elevation data stations could be found for setting a GPS in the field, e.g. the study areas I and III. With TINs, a dense network of points is required, where the land surface is complex and detailed; but where the land surface is uniformly flat or gently sloping, far fewer points must be captured. Consequently, digitizing techniques for surfacing a DEM and a TIN seems to be different and need a skill of the digitizing.

Captured data are stored as ARC/INFO *coverages*. For line *coverages*, the vertices of lines (arcs) can be converted to points with x, y, z values. From each of these points, features are converted by the TIN module of ARC/INFO into a series of connected triangles or facets. Figure 38 shows an example of a TIN feature cut out from a part of a TIN of the study area I. With ARC/INFO, there are three interpolators: linear, breakline bivariate quintic for TINs, and bilinear for lattices. In this study, both linear and quintic interpolators were tested. However, no extra *hard breaklines* were used in the interpolation process. As a result, it was found that both DEM and TIN interpolation methods trend to generate similar quality of results in this study. Since grids contain a regularly spaced, systematic sample of a surface, they are more suited than TINs for calculation of slope, aspect, sun intensity, shaded relief and cut-fill areas. Thus, TINs derived in this study were also converted to LATTICE and GRID. Figure 39 shows the flowchart of TINs creation applied in the study.

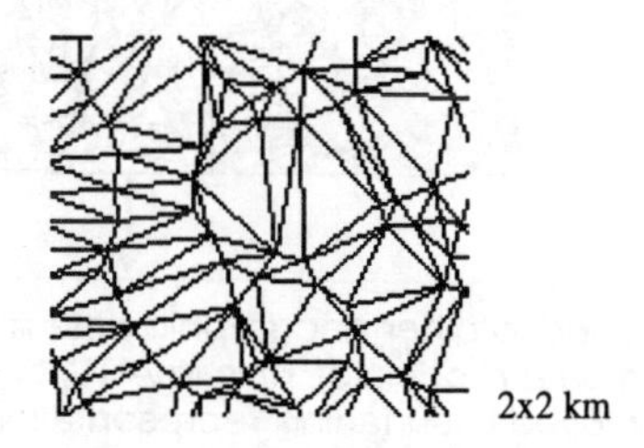

Fig. 38: Example of TIN facets subset from the TIN of study area I (Source: Author).

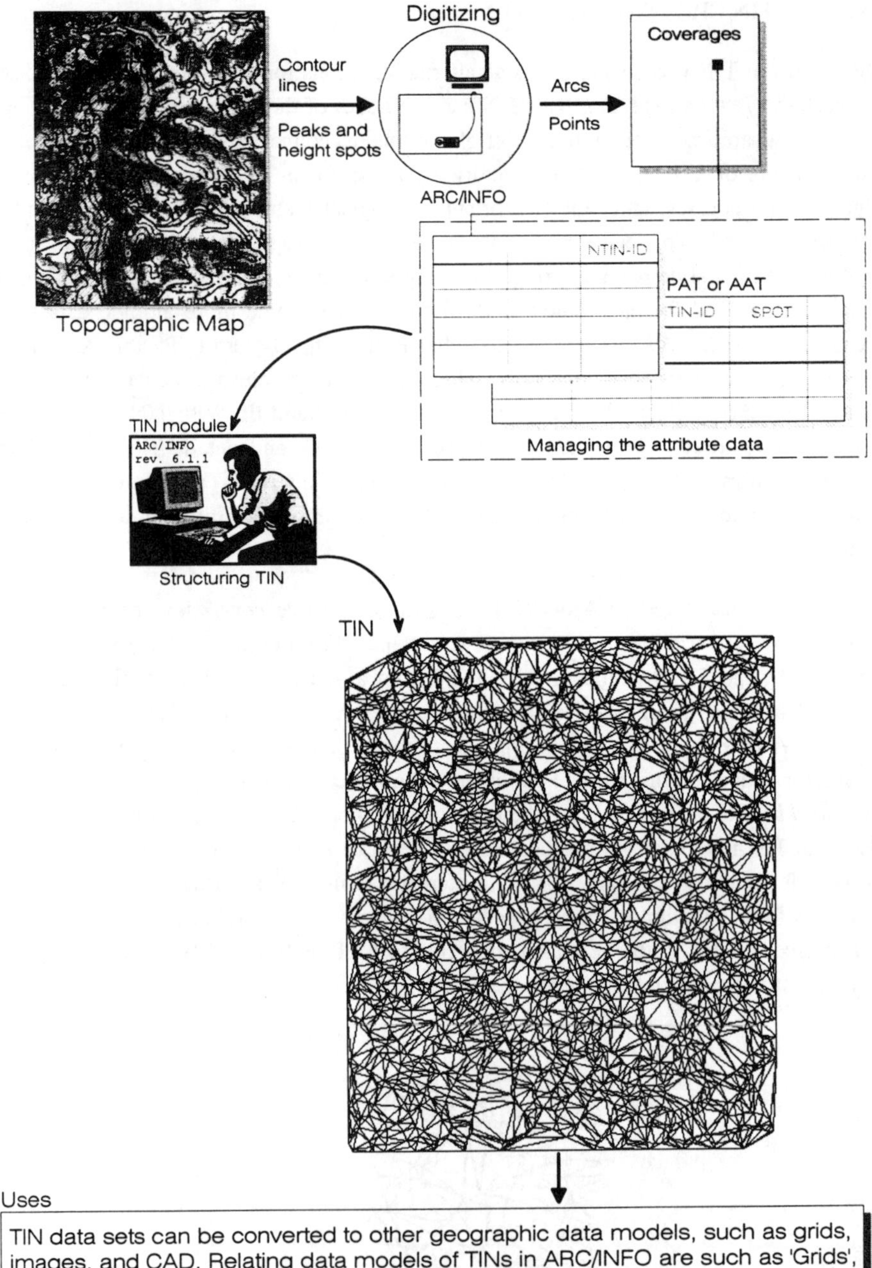

Fig. 39: **Conceptual overview of TIN structuring applied in this study (Source: Author).**

5.3.3 PRODUCTS OF **DEM** AND **TIN**

In this study, a number of software packages, such as TNTmips™, ARC/INFO®, and IDRISI™, have been adopted for the processing and the display of mathematically continuous surfaces and for the production of three dimensional (3D) perspective views. For display, the DEMs of each study area derived from ERDAS were mainly used with the TNTmips™ software. Three-dimensional (3D) perspective views of the terrain were created in different manners with different associated information, such as view angle, sun azimuth, sun elevation. A band combination of satellite images was draped over the 3D perspective view for display of resources related to terrain characteristics of each study area. At this stage, the information derived from the header file of digital satellite data are necessary for the 3D display process. With this information, the 3D display can give us a real world image of the study areas. Figure 40 is an example of a 3D perspective view subset from the DEM of the study area I. Furthermore, the DEM data files were also transported for tests in the 3D module of the IDRISI™ software. Shaded relief was also used to test the DEM and TINs products. The TIN data files were also tested and converted into a regular grid of elevations (like DEM) at 25-m spacing using the TINLATTICE and LATTICEGRID commands of the ARC/INFO.

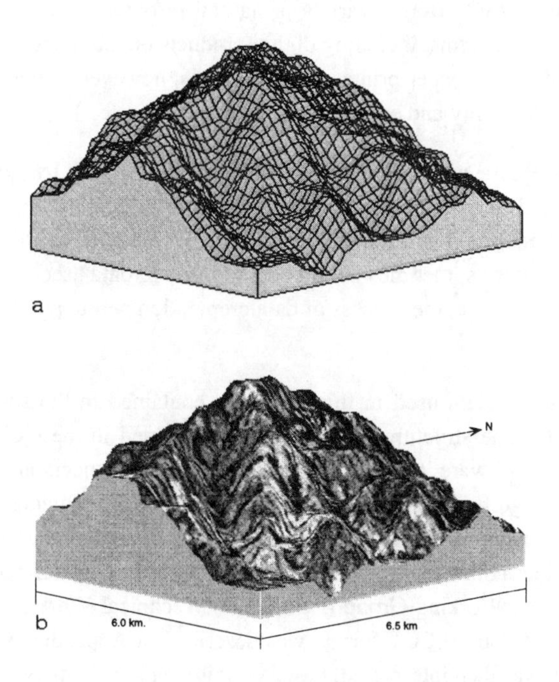

Fig. 40: **(a) 3D perspective view of a portion of the study area I displayed as a wire-frame (75x75m); and (b) Landsat TM band 3-2-1 (RGB) draped over the DEM (Source: Author).**

5.4 Satellite image processing

With the advent of higher spatial resolution for remotely sensed data, in particular as evidenced by the Landsat Thematic Mapper or SPOT, there is an increasing need to determine the most appropriate classification procedures for analyzing multispectral data. Normally, remotely sensed data can be evaluated by two methods; namely visual interpretation (analog) or digital image processing. For this study, a series of experiments involving digital image processing was undertaken for comparison. Eventually, only the better schemes were chosen and adopted.

5.4.1 DATA ACQUISITION

The main satellite data in a digital form used in this study were acquired by Landsat-5 TM. Additionally, SPOT XS was used in the study area IV in the coastal zone. In comparison to SPOT data, Landsat TM data have the advantage of a higher thematic discrimination power owing to better spectral resolution. The large coverage area of 185 x 185 km^2 against 60 x 60 km^2 of SPOT makes Landsat TM more cost effective. All data acquired were provided with support from the Thailand Remote Sensing Center (TRSC), National Research Council of Thailand (NRCT). Landsat TM or MSS, SPOT, MOS-1, ERS-1, NOAA can be ordered from the TRSC. Actually, the TRSC offers a variety of natural resources satellite data products to users world-wide in different forms, including digital products on Computer Compatible Tapes (CCT) or magnetic tape bands, paper prints, and films. It is, however, better to use the digital products owing to their flexibility and ease of distribution.

All the selected satellite data were first examined for image quality by referring to a data catalogue, quick look, and microfiche provided by TRSC. In principle, the ideal satellite data should be free of cloud and the data acquisition should meet the need of the user. In practice, these main specifications have sometimes not met the purpose of data users. Thailand has a six-month rainy season. Furthermore, the process of data preparation before purchasable commonly takes several months.

All of the acquired satellite data used in this study are contained in the so-called Computer Compatible Tape (CCT). The structure of file formats contained in the CCT product can be read from the ERDAS software (revision 7.5). However, the understanding of data file structure from a CCT is significantly important for this study. TRSC provides Landsat TM data based on the Canada Center for Remote Sensing (CCRS) ground processing systems. The Landsat TM CCT product conforms to the standard format family as defined by the Landsat Ground Stations Operators Working Group (LGSOWG), Technical Working Group (LTWG). The detail of Landsat TM data in CCT format was described in Appendix A. The meaning of image data from a CCT was then interpreted, based on this standard format.

Useful information of satellite data acquired in this study is summarized in Tab. 21. This information is necessary to use in the study. For example, upper left pixel coordinates were used to register the satellite image to the map projection. Sun elevation and sun azimuth are parameters used for 3D perspective viewing of a DEM.

5.4.2 DATA EVALUATION AND PREPARATION

Image data acquired from the Landsat satellite are raw data that have to be pre-processed. The digital data of each study areas are geocoded and stored in computer compatible tape (CCT). Geocoded products of NRCT are commonly defined in terms of the smallest rectangle falling on a one-kilometer grid unit encompassing four 1:50,000 National Topographic System (NTS) maps. These products are rotated and aligned to the UTM projection and will be provided in system-corrected or in precision-corrected form. The products used in this study are comprised of both forms. Image data used for each study area is a 1000 x 1000 pixels size, covering an area of 25 x 25 km^2 in the field.

Radiometric corrections are already done by the NRCT for each satellite data used in the study. All the satellite products used in the study are also geometrically corrected by NRCT. The level of correction methods are shown in Tab. 21. The data of study area I and II are precision-corrected products. Precision-corrected geocoded products are fully corrected (except for elevation correction with a Digital Elevation Model) in both the along-scan line and across-scan line direction by using Ground Control Points (GCPs) data. These products are rotated to a 25 meter pixel size in both directions. For study area III and IV, the data are systematic-corrected products. These satellite data are corrected for systematic errors in both the along-scan line and across-scan directions using a priori information and Payload Correction Data (PCD). These satellite data are also rotated and aligned in the UTM projection and are oversampled to 25x25 meter pixel sizes, but GCPs are not used in the correction process.

5.4.3 PRE-CLASSIFICATION

To enhance the ability of land use and land cover mapping derived from remotely sensed data, supervised and unsupervised techniques were combined in this study. All tasks concerning image processing were performed by means of computer-aided systems. Land use and land cover classification systems were prepared, based on the suitability of the use with computer-aided image processing as previously mentioned. Theoretically, most computer-assisted classifiers such as maximum-likelihood used in image classification perform class assignments based on the spectral signatures of specific pixels of image. Many classifiers do not recognize spatial patterns in the same way that the human interpreter does. Thus, a classification system should first define its land use and land cover categories based on determining spectral characteristics. At present, high spatial resolution remotely sensed data in particular from the Landsat- and SPOT- satellite, may be able to identify a few pure pixels of one object occurring on an image, such as rock outcrops, imperious surfaces, standing water. Consequently, these objects could be distinguished by computer-aided image processing, although they might not be assigned to specific land use/land cover categories in mapping. Thus, it is important to also

Geographisches Institut
der Universität Kiel

Tab. 21: Information of satellite data used in the study.

Information	Study area I	Study area II	Study area III	Study area IV	
				Scene A (left)	Scene B (right)
Product Type:	Geocoded Subscene	Geocoded Subscene	Geocoded Subscene	Geocoded Subscene	Geocoded Subscene
Product Location	Map-sheet 4946 I	Map-sheet 5038 II	Map-sheet 5442 II	Map-sheet 5433 IV	Map-sheet 5433 I
Product Scene Center Time	13/02/1992; 03:06:51	25/12/1993; 02:59:25	26/12/1990; 02:50:39	26/12/1990; 02:51:48	26/12/1990; 02:51:47
Sun Elevation at Product Center: (degrees)	41.25149	39.06972	37.42662	40.55580	40.72010
Sun Azimuth at Product Center: (degrees)	128.97227	137.63554	138.42734	136.13664	136.33931
Sensor Pointing Angle: (degrees)	0.00	0.00	0.00	0.00	0.00
Input Volume:					
Medium (tape type):	High Density	High Density	High Density	High Density	High Density
Orbit Number:	42296	52213	36267	36267	36267
Signal Acquired at:	03:05:00.0000	02:57:00.0000	02:49:00.0000	02:49:00.0000	02:49:00.0000
Signal Lost at:	03:08:00.0000	03:02:00.0000	02:53:00.0000	02:53:00.0000	02:53:00.0000
Total Number of Swaths:	81	82	84	83	84
Swaths with Sync Losses:	1	0	0	2	2
Total number of sync losses:	3	0	0	14	14
Line Length Variation:	6314.8 - 6323.1	6314.6 - 6323.2	6314.9 - 6322.8	6314.9 - 6323.7	6315.2 - 6323.7
Radiometric Options:					
Radiometric:	CAL2	CAL2	CAL2	CAL2	CAL2
Representation:	Linear	Linear	Linear	Linear	Linear
Geometric Options:					
Correction type:	Systematic	Systematic	Precision	Precision	Precision
Resampling kernel:	DAMPED 16 POINT SINC	DAMPED 16 POINT SINC	DAMPED 16 POINT SINC	DAMPED 16 POINT SINC	DAMPED 16 POINT SINC
Elevation Correction:	Not applied	Not applied	Not applied	Not applied	Not applied
Resolution (pixel size)	25 x 25 m	25 x 25 m	25 x 25 m	25 x 25 m	25 x 25 m
Number of bands	7	7	7	7	7
Data file size (pixels)	1120 x 1120	1160 x 1120	1160 x 1120	1160 x 1160	1160 x 1160
Upper Left Pixel Coordinates	578000,2101000*	634000,1632000	206000, 1854000	173000, 1384000	200000, 1384000

Remarks: * Unfortunately, it was found that the coordinates of the upper left pixels included in the header file of this image was ineptly registered. The coordinates were unusable. For use in the study, these coordinates were corrected by referencing to an old date image from the study of ONGSOMWANG (1993).

consider these features that could be delineated from the classification of the Landsat TM data.

As a result, an initially unsupervised clustering based on the iterative clustering algorithm of ERDAS was performed with original Landsat TM data during this step. According to the number of categories of the modified classification system shown in the Fig. 31, number of classes set for the clustering were started from 25 classes. Some more classes were considerably added during the classification. The initially unsupervised clustering of each study area yielded the approximate areal extent of 25 to 60 unique spectral classes. This pre-classification means to further classify the images in the next step. It can be noted that this pre-classification was based on the spectral characteristics and the natural conditions of the objects occurring on satellite images. In addition, masking technique was adopted to mask the original Landsat image of the study area III (Korat plateau) and the study area IV (Coastal zone) into two masks, i.e. land and water mask, using the result of the pre-classification. These two masks of satellite images were separately classified based on unsupervised algorithms. Vegetation index was applied to the land mask. Clustering was applied to the water mask. The results of these approaches were used in the first field survey.

5.4.4 FIRST STUDY AREA VISIT

In Thailand, both Landsat and SPOT imagery are not so quickly distributed to users until some months after data-capture owing to the data processing by TRSC. By this time, the arable crops have usually been changed and, in semi-natural habitats, signs of the management prevailing at the time of imagery have long disappeared. These problems were not underestimated in this study. The first field visit has taken place in the period between December 1993 to January 1994 to recognize the objects by their spectral features; to become familiar with the sites and to compare objects classified in preliminary work with images with their corresponding areas in the field; and to collect more information to create crop calendars.

In the meantime, optimizing land use/land cover classification schemes and determining training areas being used in classification process were done. The timing of the first field visit was based on circumstances and budget, taking into account the time of satellite data acquisition and availability. Because each of the 4 study areas was covered by a different satellite date, this field visit did not meet all the requirements set forth. To solve this problem, ancillary information from various sources together with interviews were collected in the field.

In addition, during the field work, the categories of existing land use and land cover classification systems were also examined, incorporating the initial classified images. This was to prove that which categories are corresponding to the results of the pre-classification of Landsat images. Results of this examination can be incorporated to the final optimization of land use and land cover classification system.

To complete the field work, a number of field-notes were produced as shown in Appendix B.

5.4.5 EVALUATION OF THE RESULTS FROM THE FIRST FILED VISIT

The first field survey shows significant results, necessary for further analysis in the second phase of the image classification. Tables 22 to 25 summarize the relationship between dominant land use and land cover types occurring at the study areas, the identity of spectral classes of classified images from the field, and the corresponding desired information categories from Fig. 31.

Tab. 22: Relationship between spectral classes, land use and land cover categories, and desired categories of the study area I.

Dominant land use/land cover types found in the field	Identity of spectral classes of classified images from the field	Corresponding desired information categories from Fig. 31
(1) Dry evergreen forest (With *Dipterocarp spp.*, found on ridges)	Distinguishable	Dry evergreen forest
(2) Dry evergreen forest (With various species of trees, occurring in valleys with streams)	Suppressed with shadow	Dry evergreen forest
(3) Hill evergreen forest (Found in small areas on high mountains)	Distinguishable [But similar to (1)]	Hill evergreen forest
(4) Mixed deciduous forest (Rich with big trees and dense crown closure, found in high altitude)	Distinguishable	Mixed deciduous forest
(5) Mixed deciduous forest (Disturbed, mixed with bamboo, found on lower altitude and on limestone mountains)	Distinguishable	Mixed deciduous forest
(6) Mixed deciduous (Small trees, found on lower altitude and plain)	Distinguishable	Mixed deciduous forest
(7) Dry dipterocarp (With various dominant trees)	Distinguishable but mixed to (4)	Dry deciduous forest
(8) Dry dipterocarp (With small trees, found in rolling to plain areas)	Distinguishable	Dry deciduous forest
(9) Teak forest plantation (Found on low altitude)	Mixed with (4)	Teak forest plantation
(10) Pine forest plantation (Found on high altitude)	Mixed with (4), (16)	Pine forest plantation
(11) Other species plantation (Found on high altitude)	Mixed with (4), (16)	Other forest plantation
(12) Rice fields (Found on lower lands)	Distinguishable	Paddy fields
(13) Rice fields (Found on small flooded plains near streams)	Distinguishable	Paddy fields
(14) Field crops (Found some in paddy fields, some in small areas near villages)	Distinguishable	Field crops
(15) Upland crops (Shifting cultivated areas, found on highland)	Distinguishable	Field crops
(16) Forest old clearing (Aban-doned shifting areas, covering with glass and seedling from forests)	Distinguishable	(Not directly match)

Tab. 22: (continued)

Dominant land use/land cover types found in the field	Identity of spectral classes of classified images from the field	Corresponding desired information categories from Fig. 31
(17) New forest clearing (Burned areas with bare soil)	Distinguishable only for a large area	(Not match)
(18) Villages (Group of small houses)	Suppressed by surrounding covers, but distinguishable by their spatial characteristics	Open urban
(19) Small reservoirs with dam	Distinguishable	Reservoir
(20) Asphalt roads	Distinguishable, some was suppressed by surrounding land covers	Transportation area
(21) Forest roads (Lateritic surface)	Distinguishable with spatial characteristic	Transportation area
(22) Bare soils	Distinguishable	Bare soils
(23) (Not significant)	(Pixels on deep shadow slope)	(Not match)
(24) (Not significant)	(Pixels on middle shadow slope)	(Not match)

Tab. 23: Relationship between spectral classes, land use and land cover categories, and desired categories of the study area II.

Dominant land use/land cover types found in the field	Identity of spectral classes of classified images from the field	Corresponding desired information categories from Fig. 31
(1) Rain-fed rice fields with mature rice	Distinguishable	Paddy fields
(2) Rice fields with rice in growing stage	distinguishable	Paddy fields
(3) Rice fields in planting stage (Young rice)	Distinguishable	Paddy fields
(4) Rice fields, harvested, with rice stems and bare soil	Distinguishable	Paddy fields
(5) Rice fields with rice intermixed with aquatic plants and grasses	Distinguishable	Paddy fields
(6) Rice fields, flooded	Distinguishable	Paddy fields
(7) Rice fields, in site preparation stage, moisture plowed-soil	Distinguishable	Paddy fields
(8) Kitchen garden crops, small scale cropping	Distinguishable	(Not match)
(9) Sugarcane fields in preparation stage	Distinguishable	Field crops
(10) Sugarcane fields in planting stage (Young)	Distinguishable	Field crops
(11) Sugarcane fields in mature stage	Distinguishable	Field crops
(12) Sugarcane fields, harvested	Distinguishable	Field crops
(13) Home gardens, various kinds of fruit trees and tree crops	Distinguishable	(Not match)
(14) Built-up areas	Distinguishable	Built-up areas
(15) Residential areas, usually single house surrounding with tree crops	Suppressed by (13)	(Not directly match)

Tab. 23: **(continued)**

Dominant land use/land cover types found in the field	Identity of spectral classes of classified images from the field	Corresponding desired information categories from Fig. 31
(16) Asphalt roads	Distinguishable	Transportation areas
(17) Lateritic roads	Distinguishable	Transportation areas
(18) Canals (Generally aquatic plants occurred)	Distinguishable	Applicable to streams (Not directly match)
(19) Irrigation canals with levees	Distinguishable	Applicable to streams (Not directly match)
(20) Rivers	Distinguishable	Rivers
(21) Sand mining, comprising of sand, water surface, and building	Distinguishable with consideration of turbid water surface and very bright of sand yards	(Not match)

Tab. 24: **Relationship between spectral classes, land use and land cover categories, and desired categories of the study area III.**

Dominant land use/land cover types found in the field	Identity of spectral classes of classified images from the filed	Corresponding desired information categories from Fig. 31
(1) Dry evergreen forests	Distinguishable	Dry evergreen forest
(2) Dry dipterocarp forest with rich of dominant species.	Distinguishable	Dry dipterocarp forest
(3) Dry dipterocarp forest in eastward upper-slope (With dipterocarp trees and rock outcrops)	Distinguishable	Dry dipterocarp forest
(4) Dry dipterocarp forest in lower land (With small dipterocarp trees)	Distinguishable	Dry dipterocarp forest
(5) Forest plantation (Eucalyptus)	Difficult to distinguish, mixed with (2) and (4)	Other forest plantation
(6) Shrubs and bushes	Distinguishable	(Not match)
(7) Harvested rice fields covering with single dipterocarp trees	Distinguishable	Paddy fields
(8) Harvested rice fields in alluvial flooded plain	Distinguishable	Paddy fields
(9) Harvested rice fields on foot hills, slightly rolling, rock outcrops found	Distinguishable	Paddy fields
(10) Harvested rice fields with severely eroded soil surface	Distinguishable	Paddy fields
(11) Field crops, various kinds of crops and vegetables	Distinguishable	Field crops
(12) Field crops in site preparation stage	Distinguishable	Field crops
(13) Sugarcane fields, mature	Distinguishable	Field crops
(14) Sugarcane fields in site preparation stage	Distinguishable	Field crops
(15) Upland crops	Distinguishable	Field crops
(16) Home gardens, various kinds of tree crops	Distinguishable	(Not match)
(17) Built-up areas	Distinguishable	Built-up areas
(18) Villages	Distinguishable	Open urban

Tab. 24: (continued)

Dominant land use/land cover types found in the field	Identity of spectral classes of classified images from the filed	Corresponding desired information categories from Fig. 31
(19) Reservoirs	Distinguishable	Reservoirs
(20) Ponds covering with grass and aquatic plants	Distinguishable	Ponds
(21) Dry ponds covering with grasses, wet soil	Distinguishable	Ponds
(22) Ponds with clear water	Distinguishable	Ponds
(23) Ponds with medium turbid water	Distinguishable	Ponds
(24) Ponds with very turbid water	Distinguishable	Ponds
(25) Streams	Distinguishable (With spatial features)	Streams
(26) Asphalt roads	Distinguishable (Spatial)	Transportation areas
(27) Lateritic roads	Distinguishable (Spatial)	Transportation areas
(26) Trails	Distinguishable (Spatial)	Transportation areas
(27) Exposed rocks	Distinguishable	Rock outcrops
(28) Bare soils	Distinguishable	Bare soils
(29) (Not significant)	(Pixels on deep shadow slope)	(Not match)
(30) (Not significant)	(Pixels on middle shadow slope)	(Not match)

Tab. 25: Relationship between spectral classes, land use and land cover categories, and desired categories of the study area IV.

Dominant land use/land cover types found in the field	Identity of spectral classes of classified images from the filed	Corresponding desired information categories from Fig. 31
(1) Moist evergreen forest	Distinguishable	Moist evergreen forest
(2) Dense mature mangrove *Rhizophora spp.*	Distinguishable	Mangrove forest
(3) Young mangrove *Rhizophora spp.*	Distinguishable	Mangrove forest
(4) Mangrove *Rhizophora spp.* intermixed with *Avicenia spp.*	Distinguishable	Mangrove forest
(5) Mangrove *Rhizophora spp.* intermixed with *Avicenia spp.*, trees are bigger than the type- 4 and found in high level	Distinguishable	Mangrove forest
(6) Mangrove forest with various kinds of trees species, occurring along canals where salt water is accessable	Distinguishable	Mangrove forest
(7) Disturbed mangrove forest	Distinguishable	Disturbed mangrove forest
(8) Salt meadow cordgrass, swamp	Distinguishable	(not match)
(9) Mature rubber plantation	Distinguishable	Rubber plantation
(10) Young rubber plantation	Distinguishable	Rubber plantation
(11) Orchard, pure fruit trees	Distinguishable	Orchards
(12) Orchard, various kinds of fruit trees	Distinguishable	Orchards
(13) Orchard, young fruit trees	Distinguishable	Orchards
(14) Home garden, mixed of field crops and various kinds of small tree crops	Difficult to distinguish	(Not match)
(15) Shrimp ponds	Distinguishable	Ponds (not directly match)

Tab. 25: (continued)

Dominant land use/land cover types found in the field	Identity of spectral classes of classified images from the filed	Corresponding desired information categories from Fig. 31
(16) Shrimp ponds in preparation stage, with mudflat, water was drained.	Distinguishable	Ponds (not directly match)
(17) Rainfed rice fields, harvested	Distinguishable	Paddy fields
(18) Abandoned rice fields covering with marsh cord-grass in fresh water	Distinguishable	Paddy fields
(19) Built-up areas	Distinguishable	Built-up areas
(20) Small villages	Difficult to distinguish	Villages
(21) Asphalt roads	Distinguishable (spatial)	Transportation areas
(22) Lateritic roads	Distinguishable	Transportation areas
(23) Dikes and ditches (parts of shrimp ponds)	Distinguishable	(Not match)
(24) Bare soils	Distinguishable	Bare soils
(25) Rivers with different clarity of water (different suspended sediment)	Distinguishable	Rivers, streams
(26) Canals with different clarity of water	Distinguishable	Rivers, streams
(27) Sea with different clarity of water	Distinguishable	(Not directly match)
(29) (Not significant)	(Pixels on deep shadow slope)	(Not match)
(30) (Not significant)	(Pixels on middle shadow slope)	(Not match)

For agricultural land covers, it is important to know the relationship among the cropping seasons, so that the optimum classes for discrimination can be obtained. In addition to the information from ground reference data and the existing data from the Office of Agricultural Economics (1982), the generalized crop calendars for each study area was produced as shown in Fig. 41 and 42. Using crop calendars and the date of satellite data acquisition, land uses and land covers in agricultural areas occurred in the satellite imageries could be recognized.

In comparison to the initial land use and land cover categories, there were some more new land cover types occurring during the field survey. They are corresponding to spectral classes of the initial classified maps. In addition, it was also found that not all the desired land cover units could be characterized by discrete spectral classes. Some land cover classes, recognized in the field have not been identified on the Landsat image because of their small size and their discontinuity. Some land cover units had similar signatures and could be grouped into a single spectral class. As a result, the land use and land cover units selected for supervised classification in the next step were optimized again. This classification system is shown in Fig. 43.

The knowledge and relevant results obtained from the first field survey in 1993-94 were used in the second phase of the classification with supervised algorithm. During the supervised classification, it was also necessary to consider which spectral classes could be grouped in a training process of the supervised classification, and which of the desired categories had to be subset in the training process because heterogeneous spectral features occurred in class.

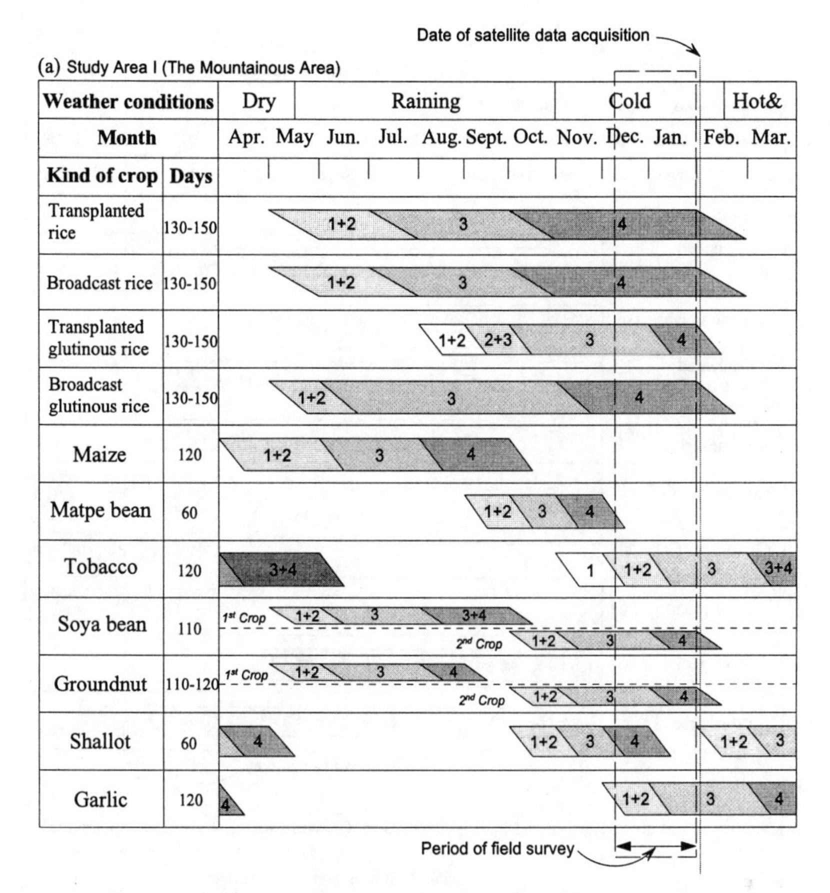

1 = Site preparation stage; 2 = Planting stage; 3 = Growing stage; 4 = Harvesting stage

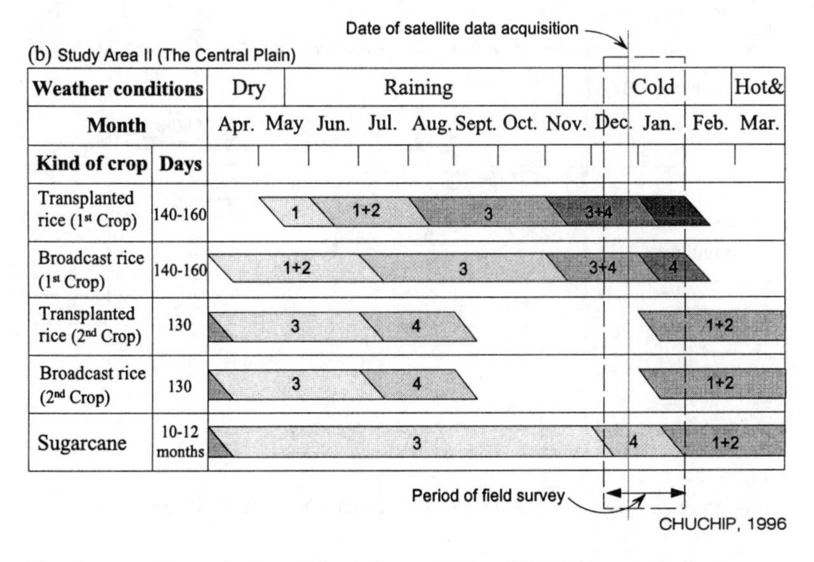

Fig. 41: **Crop calendar of the study area I (a) and II (b) (Source: Author).**

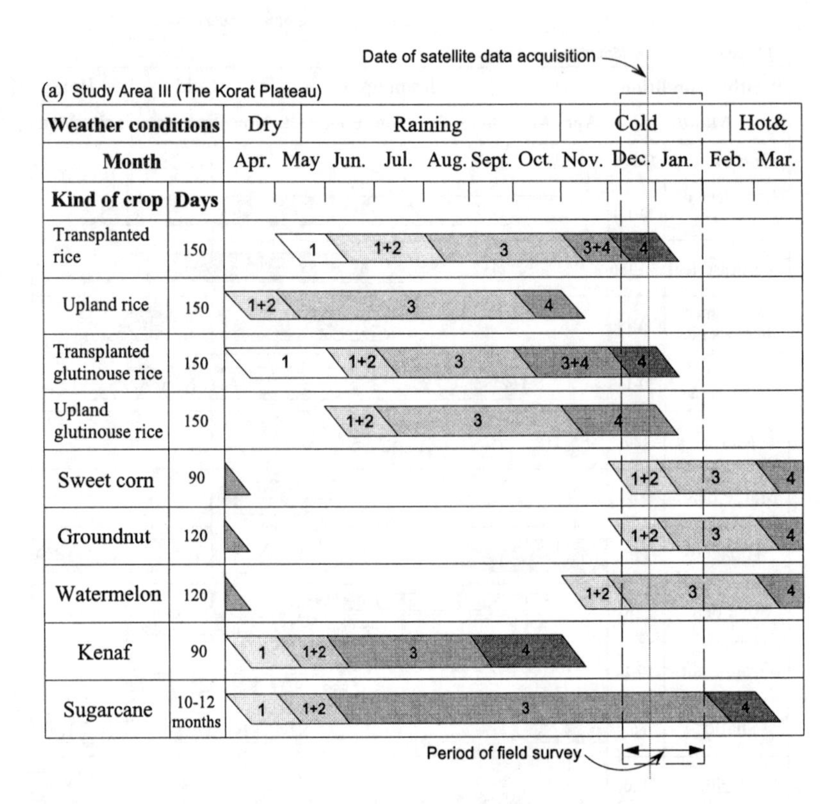

1 = Site preparation stage; 2 = Planting stage; 3 = Growing stage; 4 = Harvesting stage

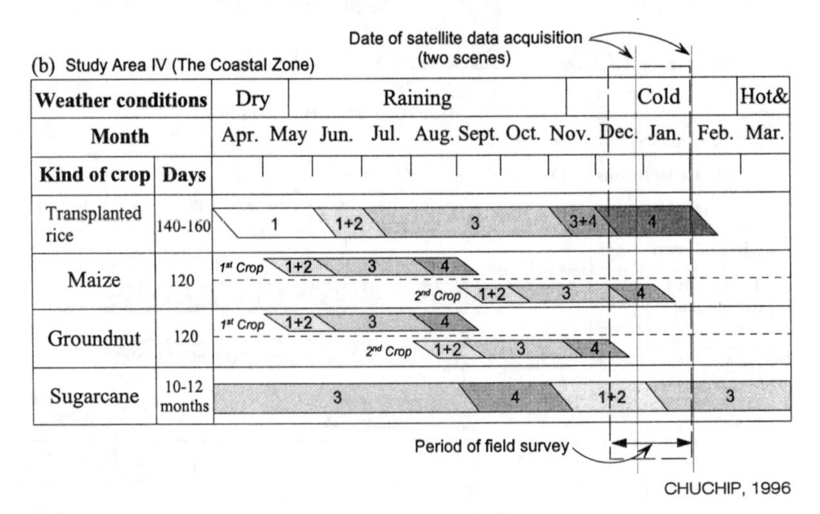

Fig. 42: Crop calendar of the study area III (a) and IV (b) (Source: Author).

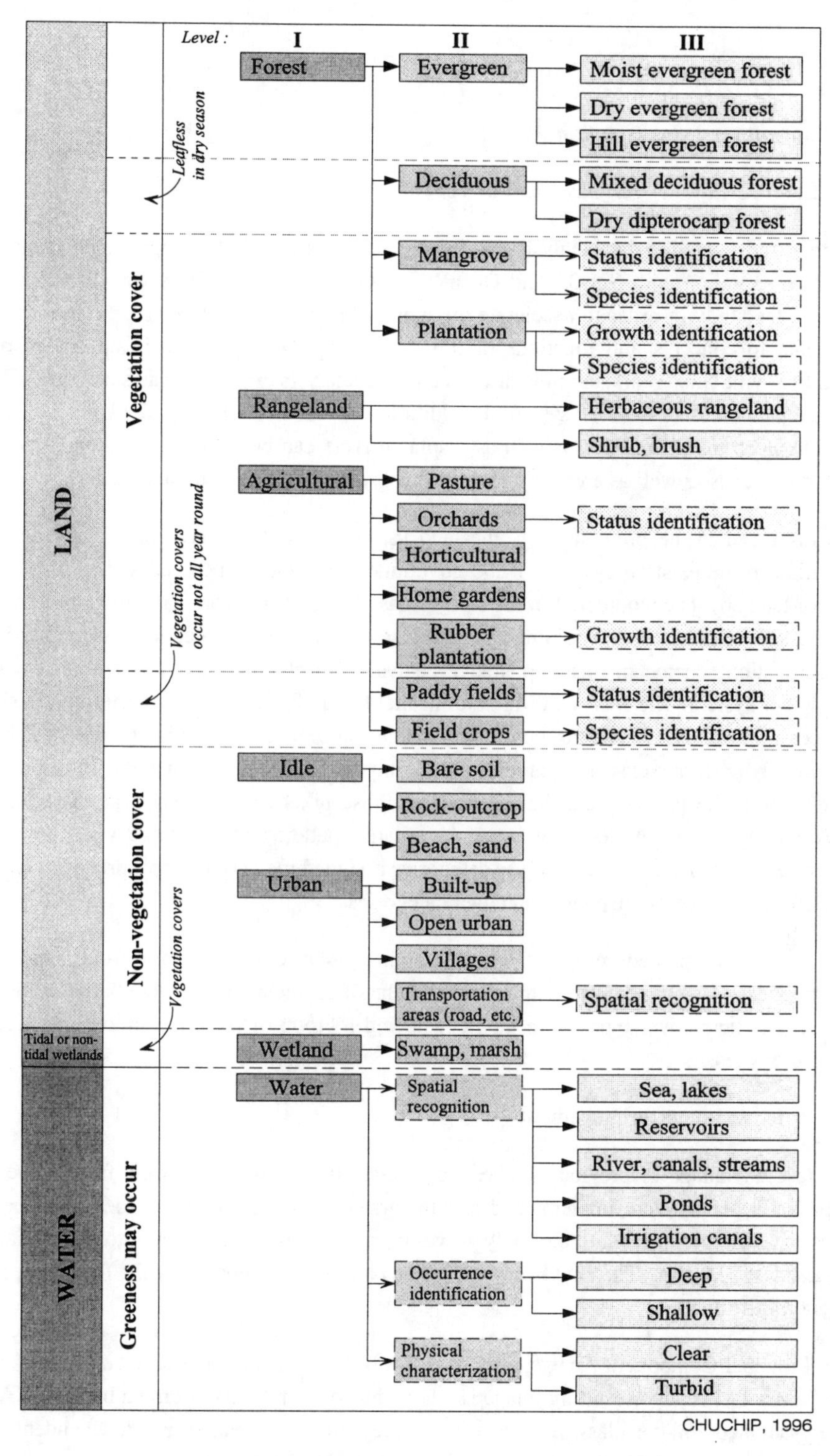

Fig. 43: **Optimized land use and land cover categories for the use in the final classification (Source: Author).**

5.4.6 CLASSIFICATION OPERATIONS

I. Conventional classification

(1) Land use/land cover classification scheme

In this study, satellite data analysis was mostly performed on the ERDAS system, running on a SUN workstation with SunOS and OpenWindows. PC-based ERDAS was also implemented during the study in order to accelerate the works. Most of the following descriptions are then relevant to concepts and functions of this software package. The main advantage of using ERDAS with Unix Workstation is the speed of the analysis process. It can save more CPU time in comparison to a PC-based system. In addition, with the multitasking of a Unix system, many ERDAS-command consoles as well as display drivers can be opened at any one-time. Issuing any commands as well as comparing image displays can be done simultaneously.

As discussed in the previous step, the land use and land cover categories were designed in addition to the existing systems of the government agencies concerned and to display all the major land covers encountered in the study areas. Some of them are naturally land cover types, while the remaining are commonly land use types. These categories were used in the classification scheme in this step. Image analysis and classification was automated with the computer system. The Bayesian Maximum Likelihood Classifier was used in the classification process. However, it should be noted in this step that this computer-assisted classifier commonly performs class assignments based only on the spectral signatures of specific pixels. It does not take into account the locations of those pixels, nor the spectral characteristics of surrounding pixels. It does not recognize spatial patterns in the same way that a human interpreter does. As a result, land use categories assigned may not be appropriate for identifying and mapping in comparison to land cover categories.

Hence, in the procedure of assigning training areas derived either from supervised or unsupervised training for use with automated classifier, most classes were based on land cover types. The land cover types were then considered a group or converted into more meaningful land use types.

(2) Satellite image enhancement

To take advantage of previous studies, some classification schemes were firstly investigated. Optimal schemes were implemented and integrated into this study. An idea of incorporating new meaningful images into the analysis was carried out by adding them into the original bands of Landsat TM data. The new bands for image data set are prepared by the following enhancement transformations:

(i) Principal Component Analysis (PCA). PCA of the original Landsat data was performed on the seven TM bands. Resulting image is a new image with 7 retransformed bands. PCA images were adopted for the classification because they are uncorrelated and independent, and are often more interpretable than the source data (JENSEN,1986).

(ii) The Normalized Difference Vegetation Index (NDVI). The expression of NDVI for Landsat TM, (band 4 - band 3)/(band 4 + band 3) was applied to create 1-band image. This band is suitable to discriminate vegetation covers and bare lands.

(iii) The expression of `band 5/band 4`. This expression is the Moisture Stress Index (MSI) for Landsat TM images. Band 4 is an indicator of biomass (*higher value = more vegetation*), and band 5 is an indicator of moisture (*lower value = more water*). Thus, a high index value indicates drier vegetation (bad), and a low index value indicates wetter vegetation (good) (ERDAS, 1991b).

(iv) The Tasseled Cap. The Kauth Transformation as shown below, was used to calculate to retransform the original Landsat TM data.

$$
\begin{aligned}
\text{Brightness Index:}\quad & 0.3037 \times (\text{Band1}) + 0.2793 \times (\text{Band 2}) + 0.4743 \times (\text{Band 3}) \\
& + 0.5585 \times (\text{Band 4}) + 0.5082 \times (\text{Band 5}) + 0.1863 \times (\text{Band 7}) \\
\text{Greenness Index:}\quad & - 0.2848 \times (\text{Band 1}) + 0.2435 \times (\text{Band 2}) - 0.5436 \times (\text{Band 3}) \\
& + 0.7243 \times (\text{Band 4}) + 0.0840 \times (\text{Band 5}) - 0.1800 \times (\text{Band 7}) \\
\text{Wetness Index:}\quad & 0.1509 \times (\text{Band 1}) + 0.1973 \times (\text{Band 2}) + 0.3279 \times (\text{Band 3}) \\
& + 0.3406 \times (\text{Band 4}) - 0.7112 \times (\text{Band 5}) - 0.4572 \times (\text{Band 7})
\end{aligned}
$$

Resulting image bands represent the *soil brightness index*, the *greenness index* (i.e. vegetation) and the *wetness index* (e.g. atmospheric scattering, haze, etc.).

(v) Band Ratios. An expression of `band 4/band 3`, `band 5/band 2`, and `band 7/band3` were used to create three new bands. These *simple band ratios* show a good discrimination of hydrological features occurring in an image. They are used to eliminate shadow effects on satellite images in some cases.

Every new band of each enhancement process was firstly used in performing unsupervised statistical signature extraction based on minimum spectral distances in assigning a cluster for each image pixel. Resulting signatures were then used to classify the image with maximum likelihood classifier decision rule. Results derived in this step were used to confirm which image sets (new bands) should be added into the original 7 bands of Landsat TM. For PCA, the correlation between the bands of resulting images was analyzed. Only the high 1 to 3 component images (depending on each study area) that contain a high percentage of the total variance were adopted for building a new image. This was because the higher components had mostly noise, such as striping and bit errors due to the high correlation. For MSI as well as NDVI, its single band was also adopted. For the tasseled cap and simple band ratios, all three bands were also adopted. 7 to 9 bands of those enhancements were combined with the original spectral bands of TM data of each study area, being used to create a new image used for further analysis. (see Fig. 44)

(3) Training samples and extracting signatures

An unsupervised non-parametric strategy (an *iterative clustering*) that incorporates contextual information was used to delineate training areas. In an iterative process, the unsupervised classification results were compared with field notes and available aerial photographs as well as maps from previous studies. In this stage, the process of clustering was iterated to satisfy the needs. Signatures of classes with best matching to desired categories were adopted to incorporate signatures derived from supervised training for use with a classifier. This unsupervised clustering was intended to be used to isolate the areas of spectral homogeneity in the image.

In addition to the cluster-generated signatures, interactively training processes with supervised approach were defined. Supervised training was mainly based on the selection of contiguous pixels or blocks of pixels from representative locations across the image as training samples. In some cases, a single-pixel training approach (seed pixel) was selected to train samples as a representative of a class comprised of groups of homogeneity pixels in a class. The selection of training pixels was aided by reference to a 1:50,000-scale topographic map, and available 1:15,000 scale aerial photographs. Training data sets were treated to locate well within the boundaries of the corresponding categories. The training data were processed statistically. Resulting spectral *signature files* containing the means, the standard deviations and the variance-covariance matrices were generated for each category and used as input to the classifier for the classification of the whole image. Since only three bands can be displayed at any time on the screen of a computer-aided classification system, an optimal selection of three bands in training process is important in order to view the meaningful data from satellite image. A series of bands combination was selected during the training process to derive the best visualization of an image. For example, TM band 4, 3 and 2 were selected to produce RGB composite images which showed best results for delineating various land cover types. Band 5, 3 and 2 were selected for viewing in the training in water surfaces on the images.

(4) Evaluating signatures

Evaluation of derived signatures created either from supervised and unsupervised training were performed prior to the application of classification decision rules. There are several built-in evaluation approaches in ERDAS, such as visually evaluating with **alarm**, evaluating **ellipse** diagrams and scatter-plots of data file values for every pairs of bands, evaluating a contingency matrix.

(5) Applying classification rule

These series of tasks, (1) to (4), were undertaken for each study area. The overall objective of an image classification procedure is to automatically categorize all pixels in an image into land cover classes. In this study, a maximum-likelihood decision rule with *Bayesian classifier* was executed using the derived signatures. Applying the classifier to classify an image has been done several times with different input parameters, such as number of image bands and standard derivation values in order to get the best result.

(6) Defining classes

After classification, a classified image is comprised of a number of classes according to the number of training samples in the signature file. Some of these classes were adopted depending upon the informational class desired. The others had to be grouped together in order to meet the informational class desired. Some spectral classes had to be assigned to cover types based on the type that dominated the class. These procedures had to be applied carefully with all of the classified images. Finally, the classified images were then evaluated by testing the accuracy of the classification.

The following flowchart shows the scheme of satellite image processing as well as the rules of making a decision used in this study.

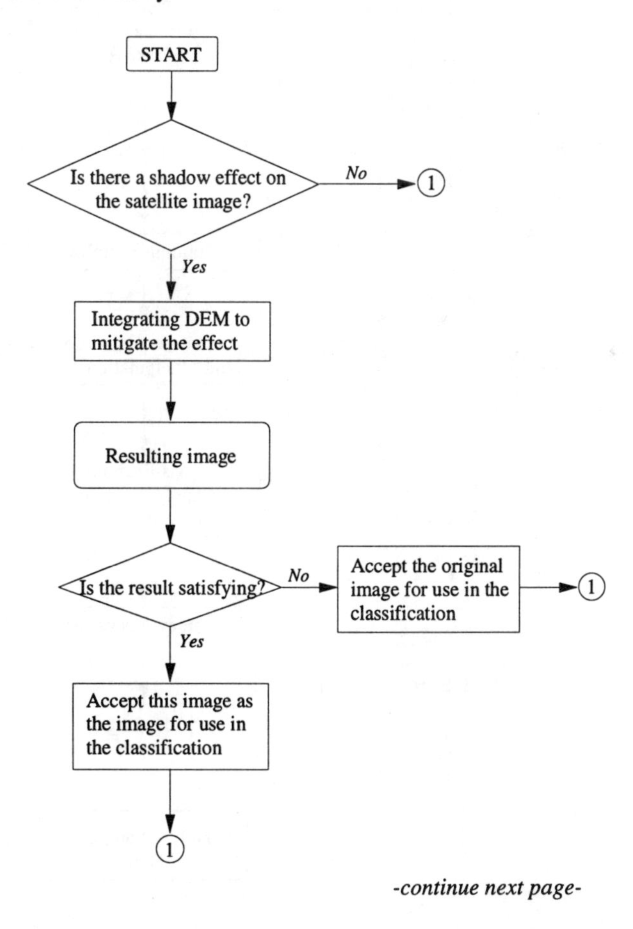

-continue next page-

Fig. 44: Conceptual flowchart of satellite image processing applied in this study (Source: Author).

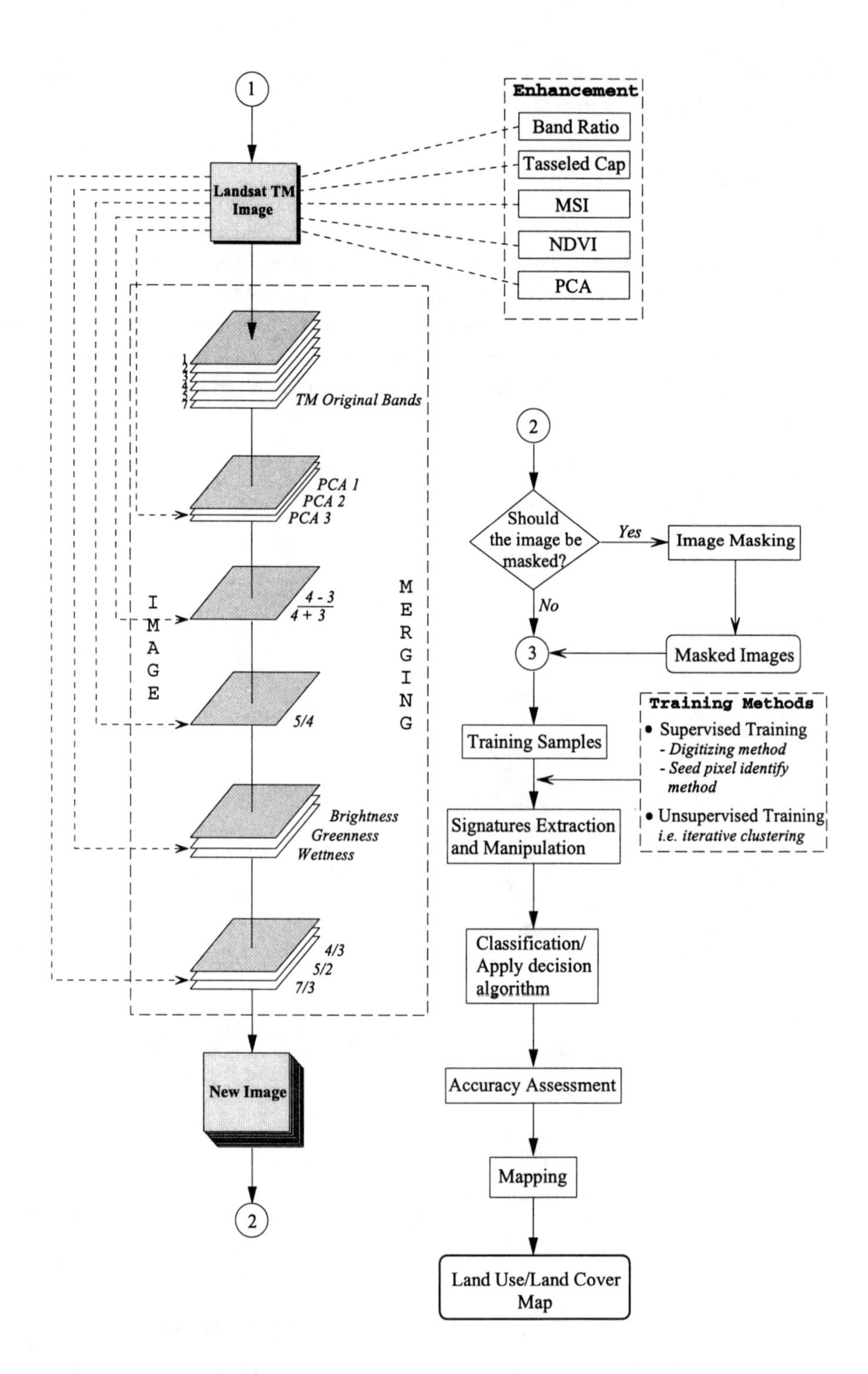

Fig. 44: **(continued)**

II. Implemented Techniques

Due to the fact that the characteristics of each image and the topography of each study area vary greatly, it is essential to implement alternative strategies for image classification. The procedures applied for each strategy are described as follows.

(1) Integrating a DEM to image classification

A DEM can produce various kinds of topographic data that can be incorporated with satellite imagery classification in various manners. A number of studies have tried to mitigate the problem of shadow effects on remotely sensed data using topographic data (as already pointed out in the literature review). Combining elevation, gradient, aspect was one of the methods found in those studies. Combining of such data can be applied both before or after classifying a remotely sensed image. For example, STRAHLER et al. (1978) initially pre-stratified a forested area into elevation ranges, and then classified the image within each stratum into various classes. HUTCHINSON (1982) classified Landsat MSS data first and then proceeded to post-classify dark pixels into shadow slopes using topographic data. A number of studies shows approaches of removing the effect of shadows on remotely sensed data by directly combining digital elevation data and remotely sensed data, such as the study of CIVCO (1989); GONG & HOWARTH (1990); NAUGLE & LASHLEE (1992).

DEM and its products were used in three stages of the image classification in this study, i.e. (i) prior to the image classification by using DEM to recognize the study area and to define training areas, (ii) prior to the image classification by direct combination into a satellite image for classification, and (iii) post-classification.

(i) Using DEM to simulate terrain models

As described above, the first field survey was done between December 1993 and January 1994 before image classification. A DEM was created for each study area. The DEM file was then subset into small pieces with rectangular shape. These small pieces of DEM files were used to simulate 3D perspective views. False colour composites of Landsat image as well as the pre-classified images were superimposed on these 3D views of the DEM to recognize the land surface looking of the study area. These 3D model images were plotted for use in the first field survey. Appendix B shows an example of a 3D model with field notes used in the field survey. It should be noticed that this study concerns with big topics that are comprised of many complex works to be done. There are 4 study areas scattered over the country. 3D perspective views with various viewing angles and directions were simulated on the computer. The study areas were recognized both from field survey and with the technology of computer simulation. This technique was applied effectively in study area I in the Ngao Demonstration Forest and study area III in the Korat Plateau that is covered with a complexity of landforms.

(ii) Combining DEM with image processing

In Landsat imagery of the mountainous areas, topographic effects are commonly contained. In areas of surfacing slopes, the response values in TM bands will be upshifted towards the brighter end of the response scale. Conversely, on away-facing slopes the response values will be downshifted towards the darker end of the response scale. Traditionally, rationing is used to reduce this topographic effect. In this study, algorithms to reduce the topographic effect are based on the study of CIVCO (1989) by integrating a derived DEM into the Landsat TM bands. Normally, Landsat data contain the information of illumination and surface material reflectance. The relationship between the sun, the spectrum of ground covers, and the topographic slope and aspect causes the incident illumination on the satellite image. In the case of study area I where is mostly located in a mountainous area, the topographic effect generally appears in darker slopes facing away from the sun and brighter sun-surfacing slopes. The unsupervised iterative clustering applied for this area did not give a satisfactory result. It was found that some land cover types were classified and grouped into improper categories. For example, Mixed Deciduous forests, covering most of the study area, in southerly slopes were classified differently to the same forest type in northerly slopes. This is because the pixels of both sides exhibit reflectance differently owing to the topographic effect. Many studies explained this effect with a similar way that the topographic effect introduced the exaggerated variance of those pixels.

A two-stages technique was employed for the removal of the topographic effects from Landsat TM data of the study area I similarly to the study of CIVCO (1989). Using the DEM generated before, a shaded relief model, also called illumination model, was created corresponding to the solar illumination conditions at the time of the Landsat overpass. Sun elevation and sun azimuth were read from Landsat data on CCTs and used to calculated this illuminuous model. The following linear transformation based on CIVCO (1989) was then to performed to derive topographically normalized images from the original Landsat TM bands, except the thermal band 6.

$$\delta DN_{\lambda ij} = DN_{\lambda ij} + (DN_{\lambda ij} \times \frac{(\mu_k - X_{ij})}{\mu_k})$$

where

$\delta DN_{\lambda ij}$ = the normalized radiance data for $pixel_{ij}$ in band λ,
$DN_{\lambda ij}$ = the *raw* radiance data for $pixel_{ij}$ in band λ,
μ_k = the mean value for the entire scaled (0,255) illumination model, and
X_{ij} = the scaled (0,255) illumination value for $pixel_{ij}$.

By visual displays of various band combinations, it was shown that this normalization was partially successful in reducing the topographic effects. To do more, the second stage of normalization was performed further. An empirically-derived calibration coefficient was determined. A number of pixels falling in both northerly- and southerly-facing slopes of the mixed deciduous forest category were selected to calculate the means and variances of spectral response for this category. The correction coefficient was then computed by the following equation:

$$C_\lambda = \frac{[(\mu_\lambda - N_\lambda)/((\mu_\lambda - N_\lambda) - (\mu_\lambda - N'_\lambda)) + (\mu_\lambda - S_\lambda)/((\mu_\lambda - S_\lambda) - (\mu_\lambda - S'_\lambda))]}{2}$$

where

C_λ = the correction coefficient for band λ,
μ_λ = the overall mean for the deciduous forest category,
N_λ = the mean on northern slopes in the uncalibrated data,
N'_λ = the mean on northern slopes *after the first stage normalization*,
S_λ = the mean on southern slopes in the uncalibrated data, and
S'_λ = the mean on southern slopes *after the first stage normalization*,

In the second stage, the calculated coefficient of each band was used to correct the topographic effect of the original Landsat TM data, also excepting the thermal band 6, by applying the following linear transformation:

$$\delta DN_{\lambda ij} = DN_{\lambda ij} + ((DN_{\lambda ij} \times (\frac{(\mu_k - X_{ij})}{\mu_k})) \times C_\lambda)$$

where

$\delta DN_{\lambda ij}$ = the normalized radiance data for $pixel_{ij}$ in band λ,
$DN_{\lambda ij}$ = the *raw* radiance data for $pixel_{ij}$ in band λ,
μ_k = the mean value for the entire scaled (0,255) illumination model, and
X_{ij} = the scaled (0,255) illumination value for $pixel_{ij}$, and
C_λ = an empirically-derived calibration coefficient for band λ.

(iii) Using DEM in the post classification stage

The incorporation of a DEM in a forest classification was applied to the post classification of the study areas II and IV. In the classification of forests, altitude has an effect on the distribution of forest types. Some range of elevations associated with particular tree species were used to stratify the study areas. For study area IV in the coastal zone, it was found that a mangrove forest class was mixed with a class defined as evergreen forest on mountains. A review of the previous study of SILAPATHONG (1992) shows that an elevation of 7.5 m (MSL) is an altitude limitation of the mangrove forest distribution. That meant that classified pixels of mangrove forest found in areas with an elevation great than 7.5 m could be defined as another forest type, here it is Moist Evergreen forest. However, there are no 7.5 m contour lines on available topographic maps. Thus, a DEM of this study area was used to interpolate the 7.5 m contour line. The simple GIS overlay using the contour 7.5 m was then used to locate miss-classified pixels. With the help of this method together with the careful consideration, the miss-classified pixels were assigned into an optimal class.

For study area III in the Korat plateau, it was found that some areas with deep-dark slopes were classified as water bodies. This miss-classification had to be corrected. It was necessary to be clear which miss-classified pixels are really the pixels on shadow slopes. Thus, the classified image was draped to the 3D surface models of the DEM. The location of the miss-classified pixels were then checked. This method could help to define and edit the miss-classified areas on image.

(2) Masking

In the study area IV in the coastal zone, there are various land use/land cover types adjacent to wetlands and streams. From preliminary spectral determination and classification it was shown that the separation of pure water pixels from wetlands and artificial shrimp ponds was not easily accomplished for Landsat TM with the supervised classification. The differentiation between shrimp ponds and water was inadequate. To solve these problems a special masking technique was applied. Unsupervised iterative clustering was performed on TM infrared bands (band 4, 5, and 7) and additional 3 bands that were derived from the band ratio transformation (mentioned above). The maximum-likelihood classification algorithm was then employed using statistical data from clustering. These operation resulted into 4 classes that could be finally defined as water class, wet soil class, vegetation cover class, and open land class. It was found that many small groups of pixels were classified as water. These pixels could be recognized as inland water bodies. Thus, optimal techniques for delineating the coastline were applied. Figure 45 shows the flowchart of the masking process used for preparing two satellite data sets, land mask and water mask. The following steps were carried out.

Firstly, except the water class, the other three categories were grouped into one class and assigned as a land class. The result of this step is an image comprised of two classes.

Secondly, filtering with boundary enhancement 3x3 mask was applied to this image to delineate boundaries between water class pixels and land class pixels. The contiguous group of pixels appearing as spatial boundaries of those classes were successfully established. The result of this stage is an image comprised of one class. As expected, a boundary between group of pixels defined as a big surface water (sea and adjacent rivers) and groups of land class and inland water class was also successfully established. In this case, these contiguous pixels can be reasonably called a coastline. It can be seen that this classified image can be divided into 2 parts, one part below a *coastline* and one part above the *coastline*.

Thirdly, with the help of the polygon fill (POLYFIL) function of the ERDAS system, the lower part was assigned to a class value (GIS value) as water class. By selecting only a 'seed' pixel somewhere in the lower part of image, pixels in this part were automatically assigned to the same class as a seed pixel (water class). Additionally, number of neighbours to search from seed pixel must be 4 in order to prevent the searching cross a diagonal vector from the seed (see Fig. 46).

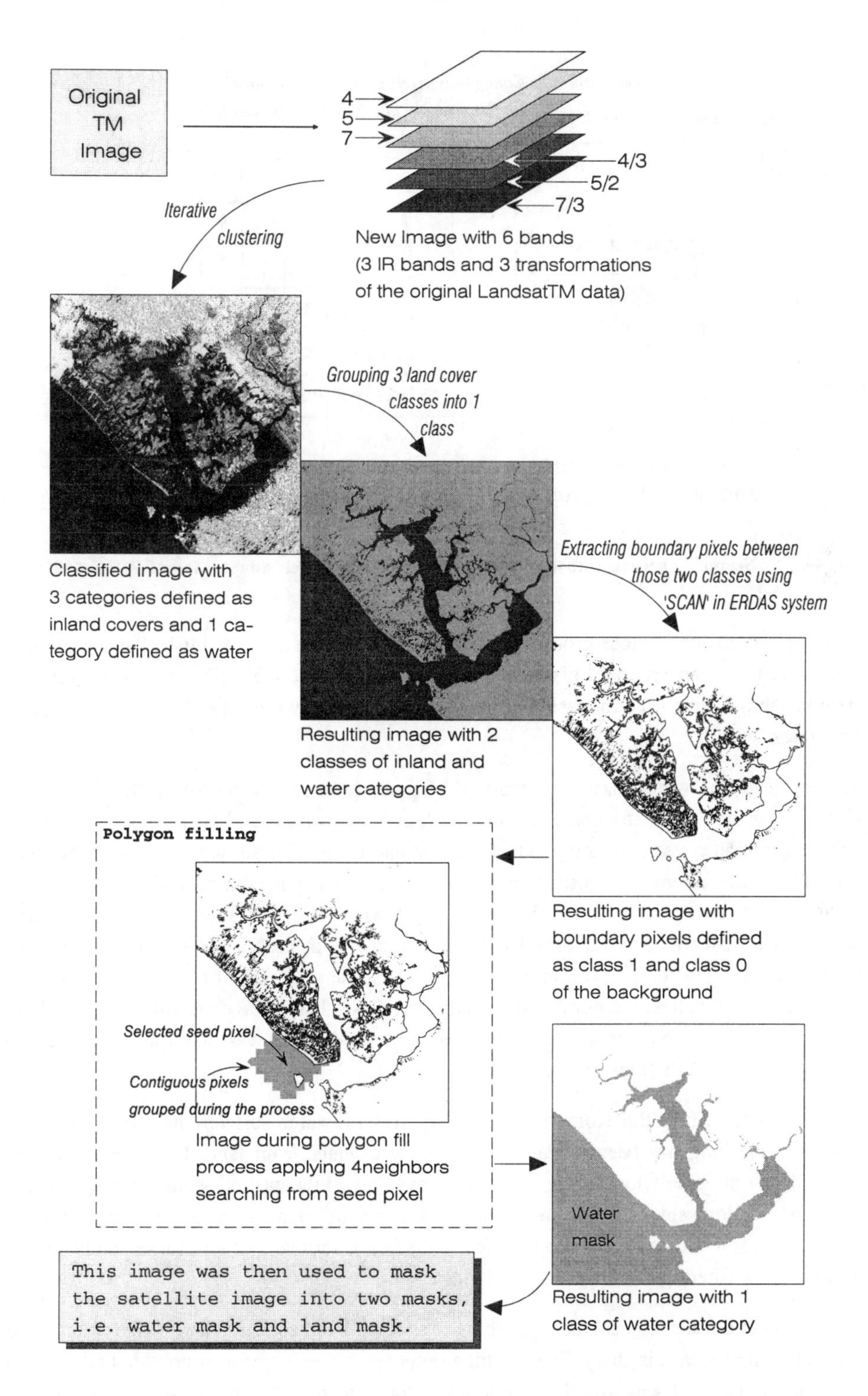

Fig. 45: **Flowchart of the masking process used in this study (Source: Author).**

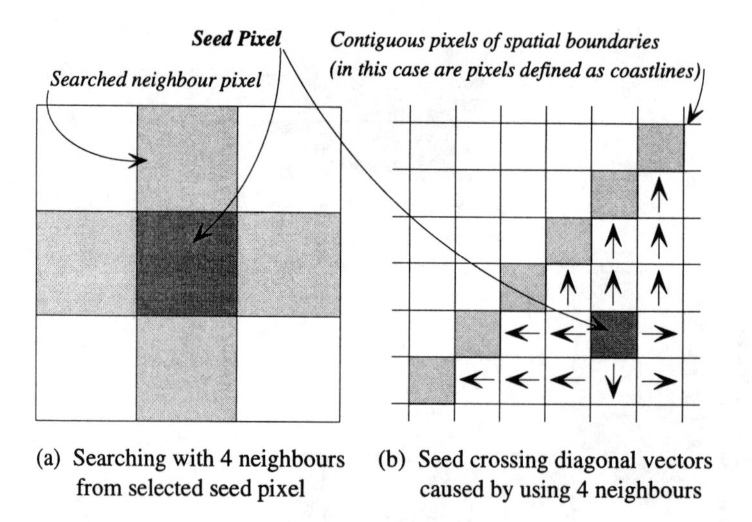

Fig. 46: Searching process with 4 neighbours from a selected pixel (Adapted from ERDAS, 1991a).

At this point, the contiguous group of pixels has been successfully established as a water class. The water class comprises all pixels below the *coastline* pixels. Finally, the water class of the final image was then used to mask the satellite imagery into two masks, i.e. a water mask and a land mask.

The image of the water mask is comprised of pure water pixels, mostly in the sea, adjacent rivers and canals while the image of land mask comprises inland water bodies, wetlands, wet soils, vegetation covers, shrimp ponds, bare land and others. These two masks were processed separately with different classification algorithms. The water mask image was classified with unsupervised iterative clustering. It seems from experience that some residual radiometric striping was observable in Landsat TM data, particularly in the visible spectral domain. This striping can be most easily seen in water areas. Thus, the selection of bands for the use in the classification as well as the number of iterative processes had to be done with a trial-and-error method in order to receive a satisfying result. The land mask was classified with the same methods as applied to other study areas.

For study area III in the Korat plateau, the pixels of water surfaces in satellite data also appeared, masking has been not applied to separate water from land. However, it was also found that some pixels located in deep water areas have low spectral values, similar to pixels on some shadow sides of the mountains. As a result, there were a few isolated pixels, producing a so-called salt-and-pepper effect. This is not serious, because these pixels could be eliminated by filtering the final classified image.

In addition to those schemes, the algorithms of spectral band selection and image enhancement were also applied in this study. The resulting output of the enhancement provided an improved view of the data and was useful for structure extraction from the image, such as roads, water bodies and their boundaries. High frequency spatial filtering was constructed to emphasize directional and non-directional differences in brightness values of the image. The enhancement image was used to identify stream locations and infrastructures such as roads, so that these features could be easily seen in the image and easily delineated by digitizing.

5.5 Ground verification and accuracy assessment

The most common means of reporting the reliability of land use and land cover maps derived from satellite data is the error or confusion matrix, also called a contingency table (CONGALTON et al., 1983). These tables are the typical result of a sampling effort in which samples of land use and land cover depicted on a classified image are compared with land use and land cover data as reference data, found at corresponding locations on the ground by means of using aerial photographs, results from field surveys, or other geodata. The sample is usually taken randomly. However, due to time and budget constraints, random sampling causes a trouble. As a result, the percent of sampling had to set to below 0.01. In some areas, some sampling units (as reference data) have not been collected or found in the field survey due to the problem of accessibility. Ancillary data, i.e. existing land use maps, aerial photographs and the reports of the departments concerned had to be additionally used in such case. The second field survey was done between September and November 1994. An error matrix for each study area was established to assess the accuracy of image classification. The procedure followed in the accuracy assessment involves the following themes.

5.5.1 SAMPLING TECHNIQUES

The initial concern in the task of assessing the accuracy of the land use and land cover classification from satellite imagery was the selection of a sample that would give reliable results applicable both to the whole classified image and to the individual land use and land cover categories. Stratified sampling based on the *main* distinguished classes was adopted. The sample to be drawn consists of a number of sampling units. Instead of selecting individual pixels as sampling units, a 3x3 window of pixels, covering 75x75 m^2 in the field, was used as a sampling unit in order to prevent sampling units to fall into 'salt and pepper' pixels. Another reason is to have sampling units with a unique ground cover and to be able to rapidly identify them in the field. Poor accessibility of the terrain in the study area I in the mountainous area and limited budgets caused the inability to use both simple random and stratified sampling. In this case, a cluster-based sampling was applied. Firstly, grids with a cell size of 40x40 pixels, 1x1 km in the field, were overlain on the classified image. These grid cells were created and registered to the image based on the UTM coordinates. Thus, the classified image was covered with 625 1-km-grid cells. Secondly, without the grid cells of the rows and columns at the boundaries of the image, each grid cell was assigned with identification numbers. There are 525 grid cells in this case. Thirdly, a grid cell was randomly selected. With this selected grid cell as a center, 8 contiguous grid cells were selected to have a group of 3x3 grid cells. This 3x3 grid-cell block was assigned as a cluster. Fourthly, simple random without replacement was done to select a number of grid-cell blocks. In short, without the grid cells in the third stage a further grid cell was randomly selected from the rest grid cells to create a new cluster. The 3rd and 4th stages were repeated until the number of clusters needed were selected. In this case, 6 clusters were randomly taken as the samples from throughout the classified image. Simple random sampling were then undertaken within all clusters. A 3x3 window of pixels was also used as a sampling unit.

Using these schemes, the field verification for this study area was successfully done. With the help of available topographic maps and the GPS, the locations of these 6 clusters were defined in the field. These schemes can solve the problem about the accessibility of this study area. Figure 47 shows the location of clusters in the image of this study area.

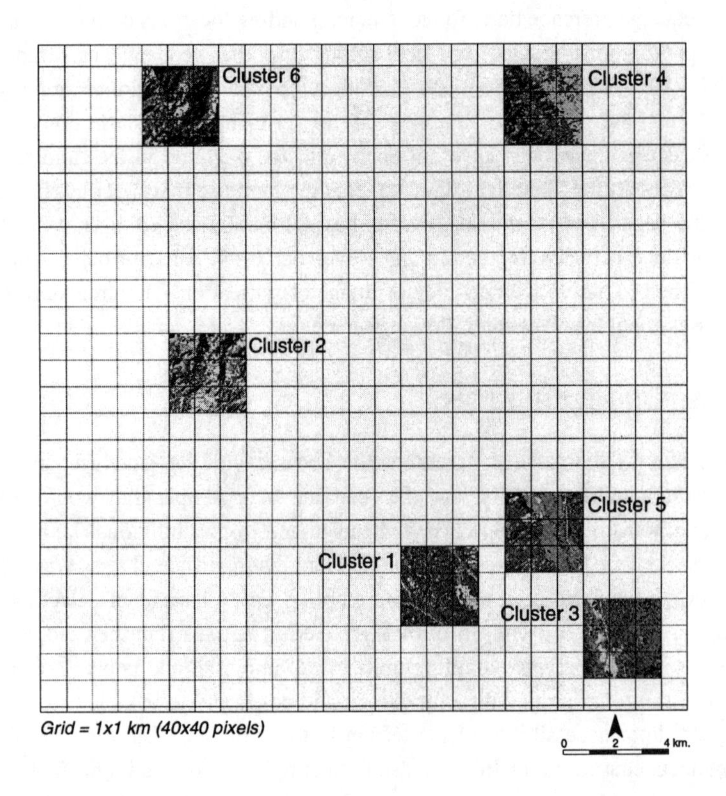

Fig. 47: Scattered clusters used in the field check applied in the study area I (Source: Author).

5.5.2 GROUND SURVEY

In each study area, the classified result from the image was compared with the same area on the ground by means of integrating them with aerial photographs and direct examination sites in the field. To make the location of sample points certain, a Global Positioning System (GPS) was used. Land use maps from previous studies were also included, since they are available in some areas. The second ground survey was done during the middle of September to by the end of November of 1994.

5.5.3 ACCURACY ASSESSMENT

(1) Classification accuracy assessment

To assess accuracy for each classified image, error matrices were established. In an error matrix, reference data were presented in rows and classified data were presented in columns. The proportion of pixels correctly classified was calculated by dividing the number of correctly classified sample position at the diagonal of the matrix by the total number of pixels checked. To measure an accuracy of individual classes, user's and producer's accuracy were calculated for each study area. In addition to the field survey, some confidential data sources, e.g. existing land use maps, aerial photographs, were incorporated into this assessment. In the case of the study area I, a land use map prepared by ONGSOMWANG (1993) and aerial photographs taken in 1983 with a scale of 1:15,000 were implemented for searching reference data. In the case of other study areas, intensive field surveys were carried out. Annual reports including relevant maps of the Office of Agricultural Economics, Bangkok, were implemented to the case of the study area II. Aerial photographs with a scale of 1:15,000, taken in 1983, were also used in the case of the study area III. A forest cover map of the Royal Forest Department, produced in 1994, and photographs taken from an airplane in 1994 were used in the case of the study area IV.

(2) DEM accuracy assessment

The accuracy of DEMs and their derived products are of crucial importance because errors in the base data will propagate through spatial analyses. However, the accuracy of DEM's products, i.e. aspects and slopes, have not been evaluated this time owing to lack of reference data for comparative use. There are no comparable elevation data for this assessment, except the values derived from topographic maps. Evaluations using data from these maps would be bias because DEMs constructed in this study are also based on terrain data derived from those topographic maps. However, simple accuracy assessment schemes were tried out.

To characterize the DEM data quality, a number of grey-level stretches were computed. The quality of DEMs display were assessed visually. This included reasonableness, conformance to general knowledge of terrain shape, and geomorphic consistency. Shaded relief and 3D perspective using the DEMs were draped with streamlines and water bodies to prove the smoothness and the correlation between them.

5.6 Land use and land cover classification system optimizing for use with satellite imagery analysis

As stated in Section 5.1.1, the classification systems used in this study were modified from previous systems and also based on computer-aided classification. After the satellite data analysis was successfully carried out, classification results were evaluated. It showed that land use and land cover categories were closer to being nearly land cover types than land use types. The classification results shows that it is better to convert the land cover categories of classification systems from the former phase into more meaningful land use categories. Furthermore, the land use and land cover classification systems of all study areas should be optimized and reorganized again.

5.7 Computer mapping systems

Computerized plotting seems to be the easiest and most flexible way in cartographic mapping nowadays, if the facilities are available. In this study, modern approaches to digital satellite image map production were entirely based on a computer-aided mapping techniques. The data transfer between each step was based on computer-aided system by means of digital-to-digital conversion of the data. A large format high resolution raster plotter system was used to produce land use and land cover maps (Appendix E).

The ERDAS system was used in this study only for image processing while the ARC/INFO system was used for geographic information analysis. None of them have been used for mapping. Computer Aided Design (CAD) systems and illustration software systems were adopted as other choices for producing thematic maps. AutoCAD was adopted here as the CAD system. Micrografx Designer was adopted as an illustration software system.

Since digital data in a graphic file format should be viewed and modified at any time on various computer mapping systems, several mapping software vendors always provide the feasibility of storing, retrieving and transferring digital data in standard data formats. Image data derived from the ERDAS system can be converted for graphical data exchange. The Tag Image File Format (TIF) is one of a few formats that the ERDAS 7.5 supports. Since TIF has become widely used in many PC-based software packages involving graphical works, thus, the TIF format was mainly used with raster data for transferring between different systems in this study. In terms of the vector data exchange, the Drawing Exchange File (DXF) format was a standard one to be used with vector data derived from the ARC/INFO in this study. In addition, the replaceable Windows Metafile (WMF) was also adopted in some cases.

In essence, CAD systems handle geographic data in the same manner as topographic map. CADs have also been used for the production of topographic maps. The CAD system by itself, however, could not automatically shade each parcel based on values stored in the database of a GIS (e.g. an attribute table of ARC/INFO). As a result, an interactive manual legend was necessary in mapping with a CAD system in this case. In this study, the CAD system has provided valuable capability to quickly generate base maps for use in the field surveys.

In short, the AutoCAD software package was adopted to import DXF-files from ARC/INFO. These drawing data were then manipulated and added with symbols, and attribute data. The graphical elements (lines, symbols, etc.) in the drawing data were set into layers by means of the AutoCAD. These drawing data were then exported from the AutoCAD in DXF-file format to an illustration software package. Micrografx Designer software package was selected to produce thematic maps owing to its capabilities in reading DXF files error-free including layers as well as TIF files, the capabilities in manipulating raster and vector data at one time, and in plotting a large format map such as DIN A0. Some of artworks in forms of thematic maps provided from the Micrografx Designer can be found in Appendix E.

6 RESULTS AND DISCUSSION

6.1 Evaluating the procedures and products of surface modeling

6.1.1 EVALUATION OF DIGITAL ELEVATION MODELS

The modeling of a digital raster surface, grid-based DEM, using terrain data from a topographic map by means of ERDAS concept yields satisfying results with some limitations. It was found that the quality of a DEM surface was depending on four factors, i.e. weighting function, blocking factor, search radius, and the manner of terrain data points distribution. Figure 48 (a) and (b) show the portions of a resulting DEM and its products derived from the case of the study area I (see also Appendix C). Not all built-in weighting functions are suited for modeling a surface using topographic data. Based on the visual appearance, the results derived from the weighting function 7 show an accurate-looking topographic surface both in the manner of shaded relief and 3D perspective view. This derived DEM may be the best choice for integrating into the Landsat TM data in the procedure of the topographic effect reduction. However, this function is relatively complex and takes CPU time. The weighting function 3, 4, 5 and 6 give also acceptable results. The visualized quality of interpolated contours derived from the DEM did not so vary in the function of 3, 5, 6 and 11. Resulting contour lines are relatively smooth. Conversely, the contour lines are relatively angular in the case of the function 1 and 7, but it appears looking accurate. Some models, i.e. 2 and 8, contain an error of contour interpolation, most of which appears as small closed contours. Results from other study areas behave also in the same manner.

Blocking factor and search radius also effect resulting DEM. In any case, these values used variably in each study area due to the vary density of topographic data used in surfacing. Unfortunately, no good reason was found to state which of the blocking factor or search radius should be used. Thus, a trial-and-error process is still required to determine these parameters. Table 26 shows the values used in this study.

Tab. 26: Blocking factors and Search radius used in each study area (Source: Author).

Study area	Blocking factor	No. of point/Block	Search radius
I Mountainous area	875	127 (74%)	875
II Korat plateau	1408	166 (97%)	1400
III Central plain	2612	3 (1%)	2610
IV Coastal zone	1500	170 (100%)	1500

Due to the fact that the surface program is critical with the number of terrain data point in each block, not more than 170 points and at least one terrain data point should be included within each block. Surfacing a terrain are difficult in an area where comprises both high mountains and plains, such as in the case of the study area III and IV. Derived terrain data are dense in

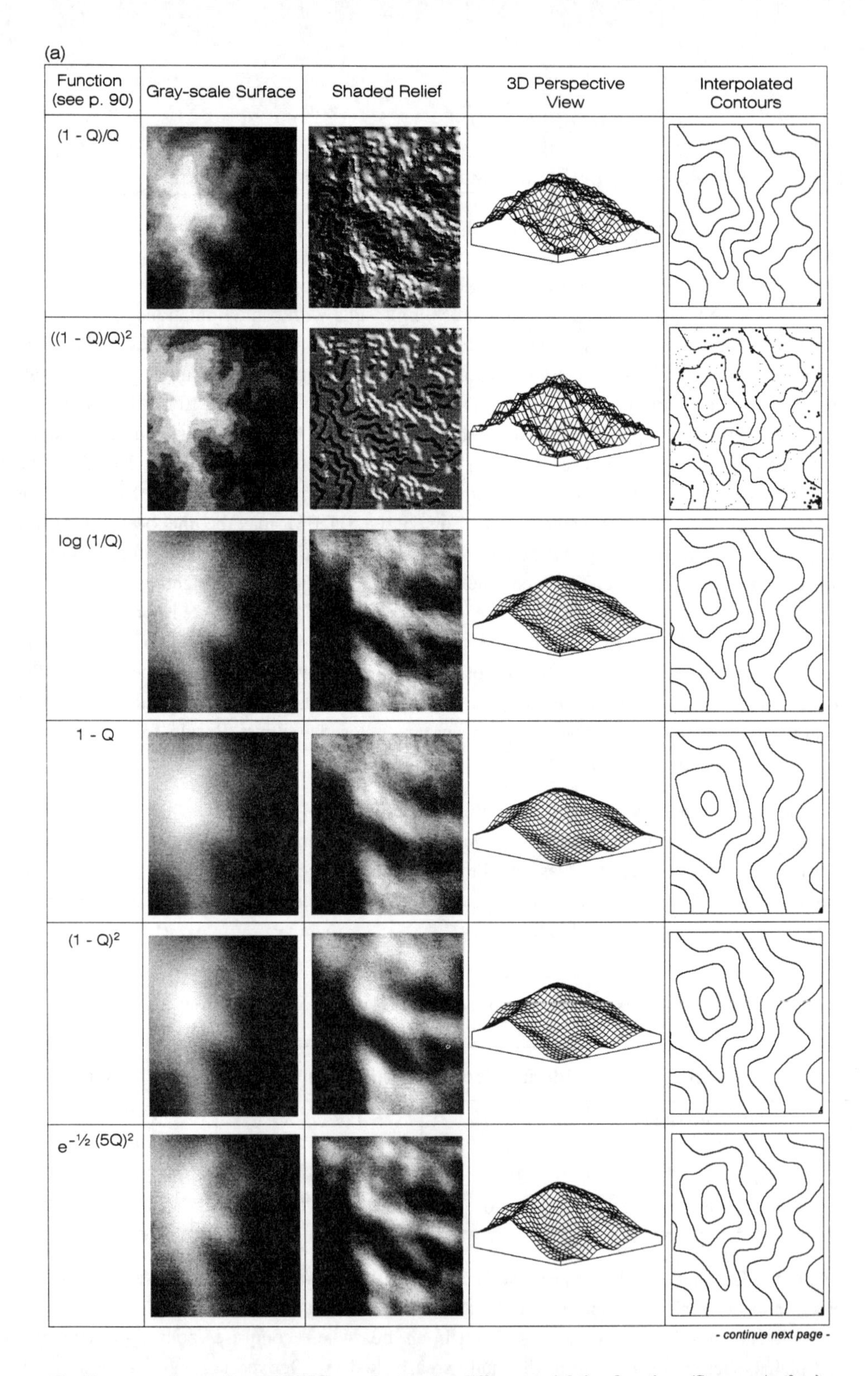

Fig. 48: Comparison of DEM's products using different weighting functions (Source: Author).

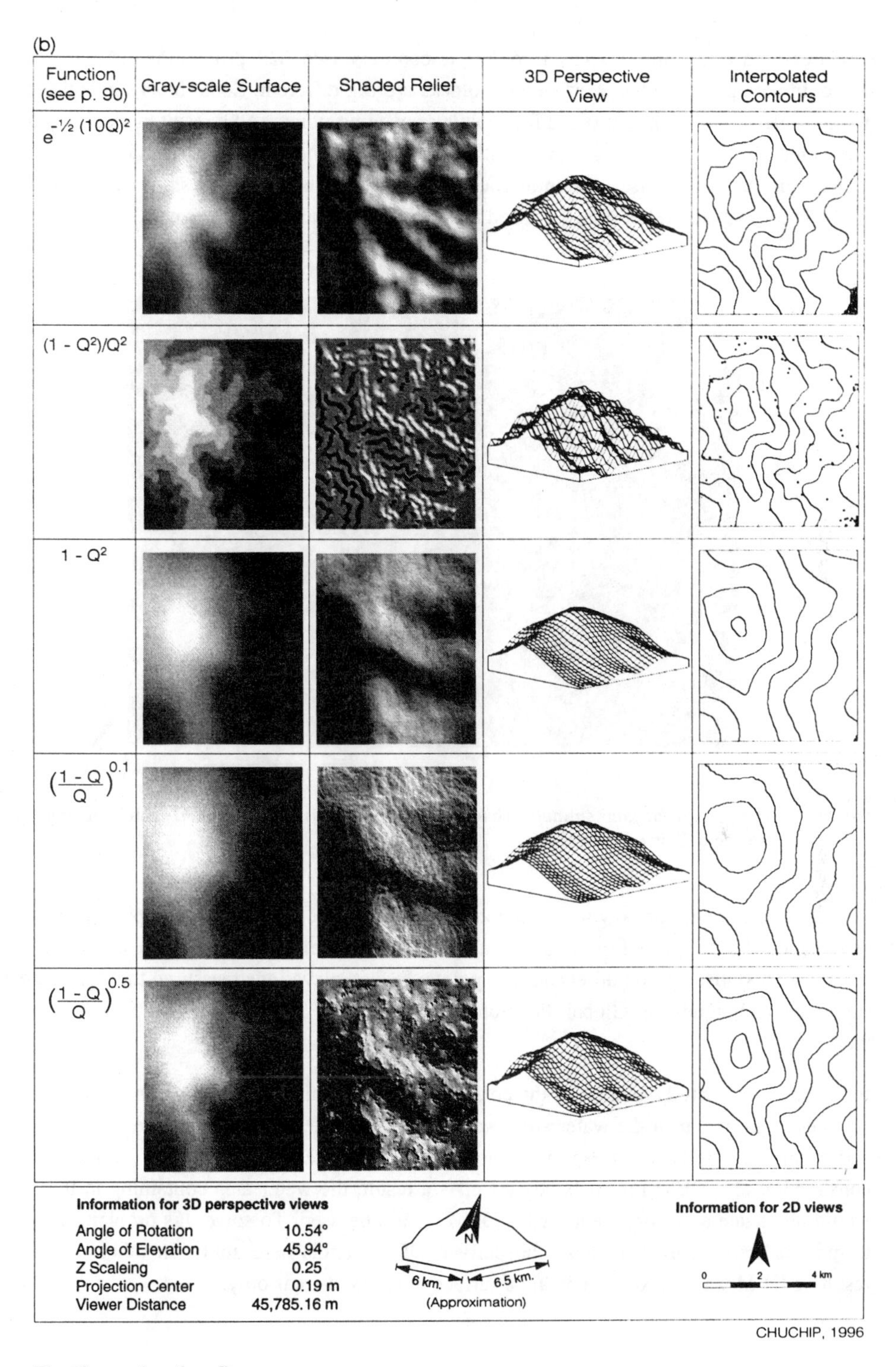

Fig. 48: (continued)

high areas and sparse in flat areas according to contours used. The problem is that the number of terrain data points fallen in processing blocks are big different. Some blocks contain many points while some blocks contain a few points or even no points. As a result, a resulting DEM may contain *no-data surfaces,* pixels with a zero value, and sometimes contain a jagged or stepped surface. This causes discontinuities in DEM products. Figure 49 shows an example of discontinuous shaded relief image derived from a DEM containing such errors.

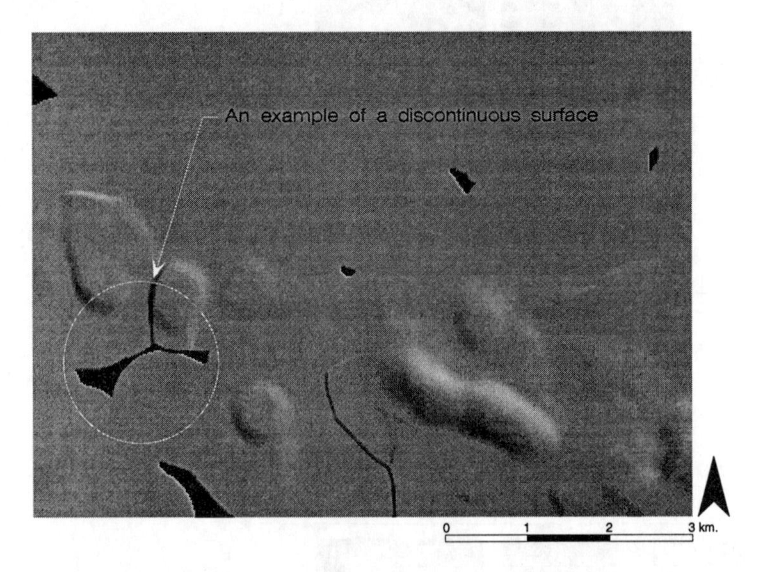

Fig. 49: **An example of a discontinuous shaded relief image derived from a DEM containing no-data surfaces (Source: Author).**

In order to accurately portray the variation in terrain surface, optimal topographic data should be included. Techniques of digitizing contour lines and other topographic data should be done to have a good distribution of terrain data points. In fact, elevations measured through standard surveying techniques or Global Positioning Systems (GPS) could be included in surface modeling.

In the case of the study area IV in the coastal zone, the resulting DEM contains many *no-data* surfaces in particular in the water area. Actually, a similar terrain value can be applied for the water surface. However, there were no terrain data used for the water surface in the construction of a DEM for this study area. As a result, the water area containing in the DEM would propagate an error whenever the DEM should be used. To solve this problem, masking techniques were applied to mask the derived DEM in this case study. In other words, the resulting DEM was masked to contain a surface of the land areas only.

6.1.2 EVALUATION OF **TINs**

Structuring TIN was also done using digitized contours and height spots. Breaklines, surface discontinuities and no-data areas were used in some cases. In the case of the study area I, resulting TIN appears very complex due to the triangular facets constructed from digitized contours. As mentioned before, the topography of this area comprises complex mountains. The topographic map of the area contains dense contours. In the first try, derived TIN comprised many complexity of triangular facets as well as many flat triangles. To mitigate this problem, weeding was used to remove *vertices* from digitized contours before building a TIN. The resulting TIN of the study III appears as good terrain representation after the scheme in the case of the study area I has been followed.

In the case of the study area II in the Central plain, the construction of a TIN was also tested. In this case, only existing height spots containing in the topographic map have been used, due to the lack of other terrain data. Existing water bodies have been not used as *breaklines* in this time because it was not intended to use the resulting TIN in the further step. However, the resulting TIN shows a good appearance. It can be noted that in flat plain area, a TIN can be constructed using existing measured points. The resulting TIN can be used to interpolate contours that are essential for any purposes, such as for the use in planning the area intensively. However, a DEM was used to interpolate contours instead of using the TIN for this study area because contours were needed in raster forms for further use in another part of the study (Section 6.2.1). The DEM of this study area was used to interpolate raster contours with an interval of 0.2 meter. In fact, there have no 0.2 m contour interval on available topographic maps. These contours were implemented in order to interpret some land use types occurring in the image of this study area.

TIN surface modeling concretely shows a better result than grid-based modeling of DEM in the case of the study area IV in the coastal zone. In this case, the shorelines were used as a *hard breakline*. Actually, the water surface can be treated as a flat area with a constant elevation. In this case, the water surface was treated as *no-data* area because of the inland data interest in this step. The surface model for water areas was also constructed by modifying the method of modeling a DEM and a TIN. The *breakline* was used here to accommodate and to control interpolation algorithms for generating contours. The interpolation algorithms can recognize the shorelines while restricting contours from passing through the water bodies. In addition, a water mask polygon was used as a *hard ERASE* feature in the process of creating the TIN because the interesting area is comprised of separate land surfaces, i.e. main land and two islands. The resulting TIN is contained in the non-convex hull which is essential to prevent the generation of erroneous information in regions of the TIN outside the actual data set, i.e. water area. At first, any ERASE features were not applied. As a result, the resulting TIN showed an invalid surface representation, as shown in Fig. 50. It was found that these invalid surfaces caused an invalid interpolation.

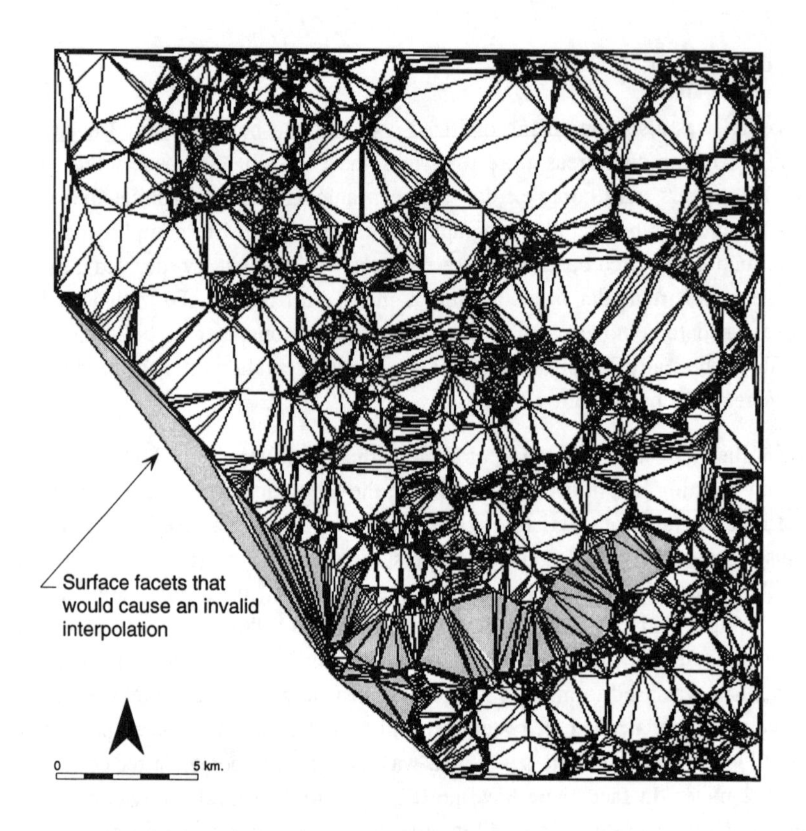

Fig. 50: **Example of invalid surface representations derived in the case of the study area IV before applying ERASE features (Source: Author).**

6.1.3 PRODUCTS AND ACCURACY OF **DEM**s AND **TIN**s

The resulting DEMs exist in form of raster images stored as LAN files of the ERDAS software. These image data arc arranged in a Band Interleaved by Line (BIL) format. The pixel values are 16-bit packing with signed data. Each DEM file is already included an important information, i.e. the type of map unit in an integer number, the number of area units (hectares) in a real number, the map coordinates in a real number, the pixel size in real numbers. Thus, they can be used without complexities. Tests were undertaken to transfer these DEM files to some different software platforms, i.e. TNTmips, IDRISI. No problem was found in further use. All DEM files were tested to interpolate some relevant products in order to make sure that their products can be used in the further step of this study. Samples of the DEM's products are shown in Fig. 51 to 54. In view of the results presented in these figures, they appear to be useful for support in further studies. (see also Appendix C for more descriptions of three-dimension viewing)

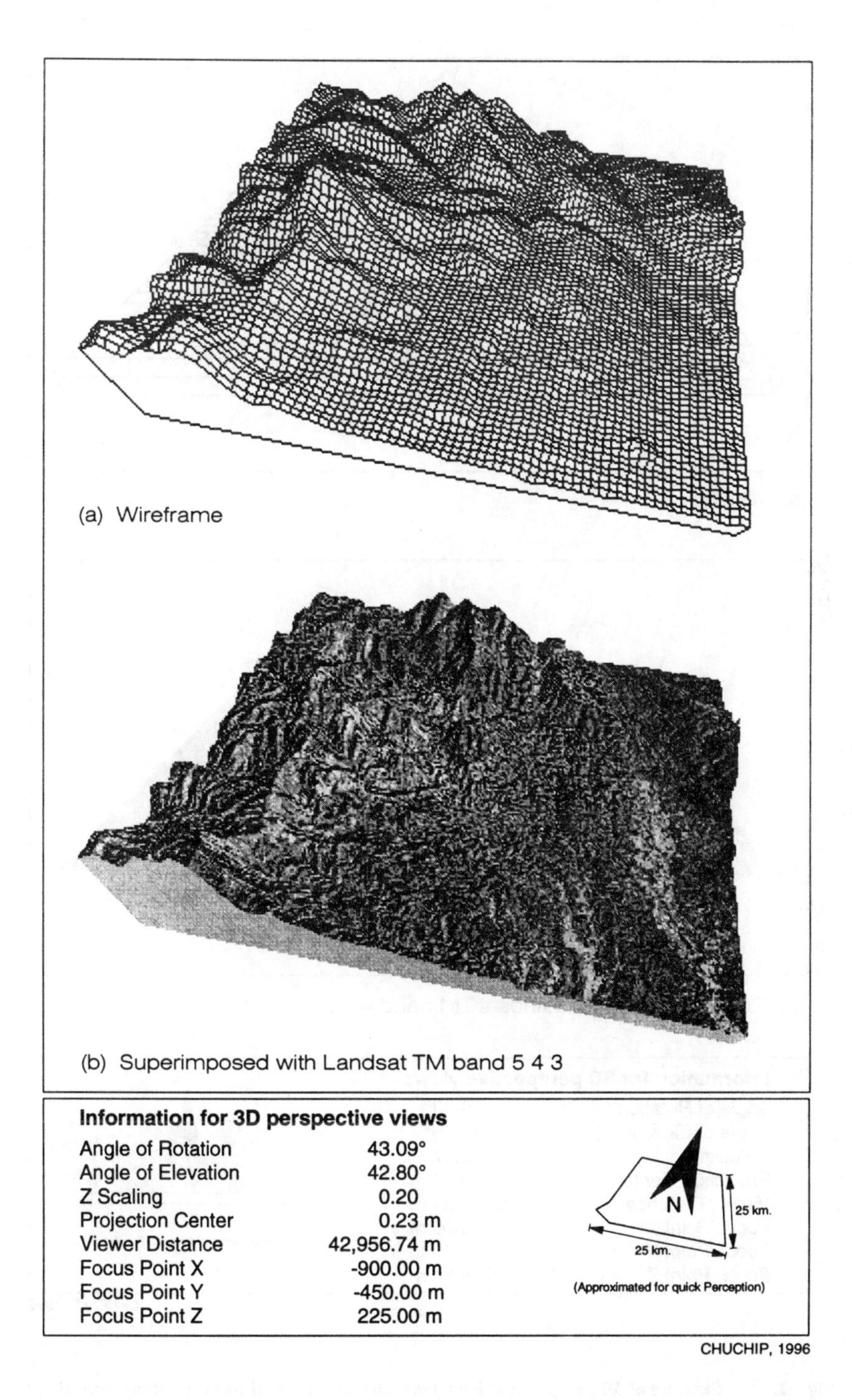

Fig. 51: **Samples of 3D perspective views generated from the DEM of the study area I (Source: Author).**

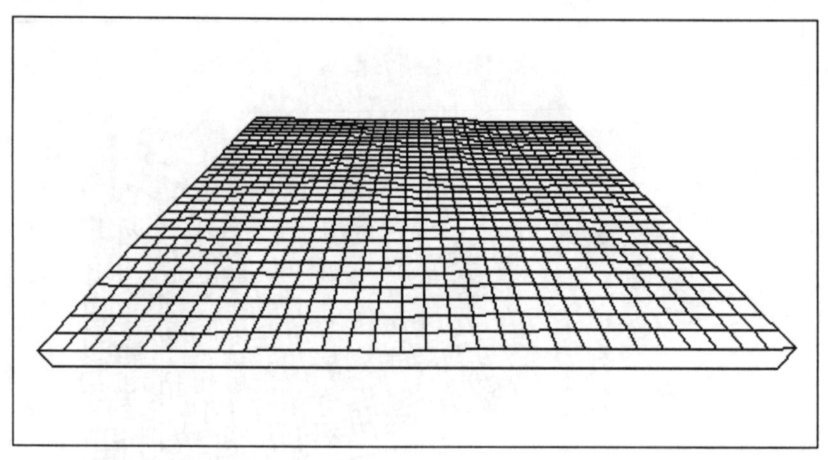

(a) Wireframe

(b) Superimposed with Landsat TM band 4 3 2

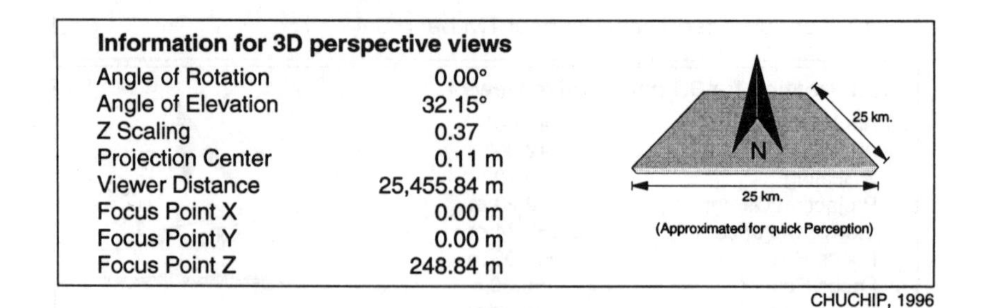

CHUCHIP, 1996

Fig. 52: Samples of 3D perspective views generated from the DEM of the study area II (Source: Author).

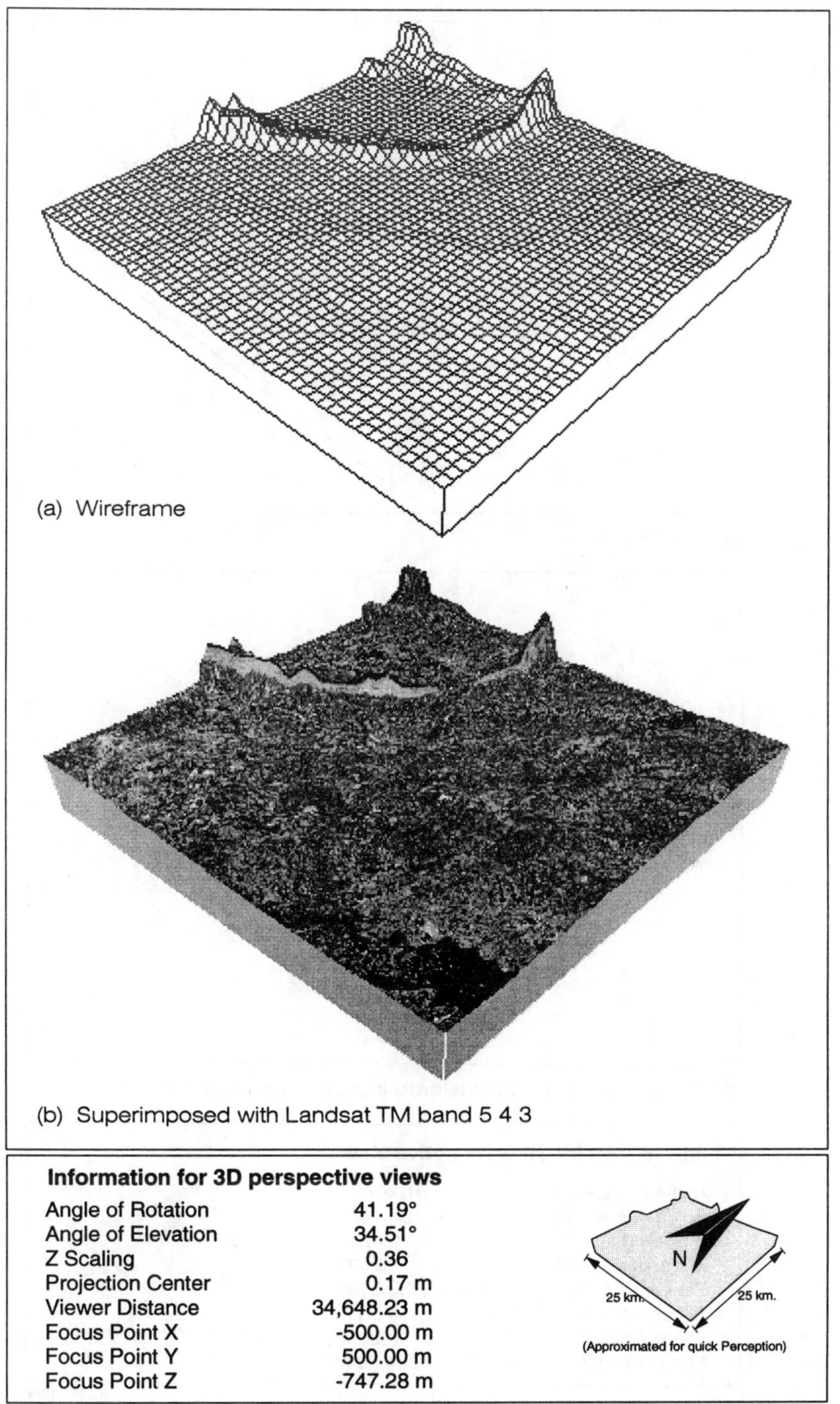

Fig. 53: Samples of 3D perspective views generated from the DEM of the study area III (Source: Author).

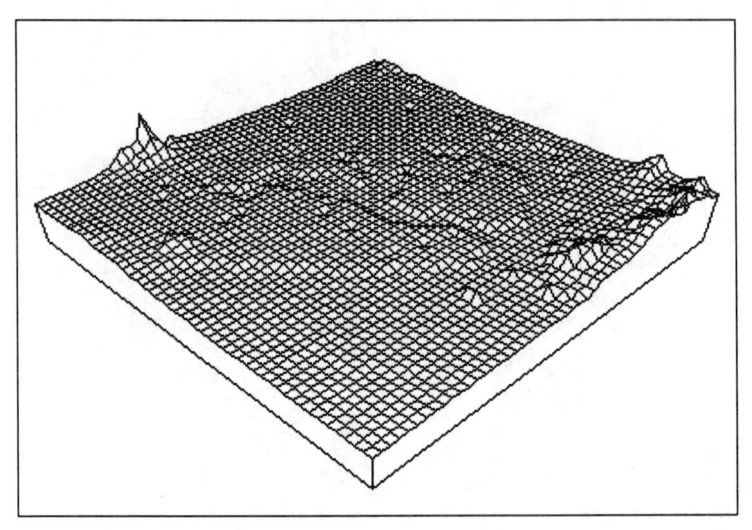

(a) Wireframe

(b) Superimposed with clustered image of Landsat TM

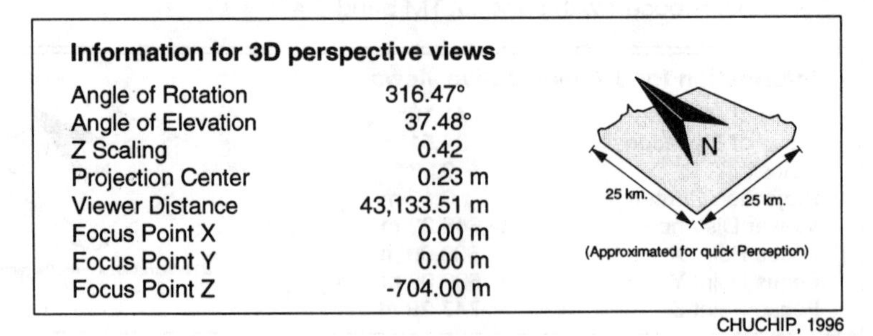

Information for 3D perspective views

Angle of Rotation	316.47°
Angle of Elevation	37.48°
Z Scaling	0.42
Projection Center	0.23 m
Viewer Distance	43,133.51 m
Focus Point X	0.00 m
Focus Point Y	0.00 m
Focus Point Z	-704.00 m

CHUCHIP, 1996

Fig. 54: **Samples of 3D perspective views generated from the DEM of the study area IV (Source: Author).**

While a resulting DEM is a systematically-spaced sample of surface elevation measures, a resulting TIN is comprised of facets that are a connected set of irregular-spaced x-y-z locations. A TIN is stored in the form of ***tin coverages*** by means of the ARC/INFO. Each resulting TIN can be retransformed to a LATTICE that is a regularly-spaced sample of points representing a surface. That means a TIN can be also transformed on grid-based data sets as a DEM. Figures 55, 56 and 57 show the resulting TIN created in this study. Using TINs to simulate a 3D view provides also a good surface view.

Both DEM and TIN data have been tested to generate slope maps as well as to construct contour maps. However, there have been no intensive assessment of their accuracy in this study. It can be, however, noted that the relative accuracy of both DEM and TIN is dependent on the accuracy of topographic maps used. It is the estimation from estimated terrain data points (data from topographic maps). Surface modeling with TIN™ shows its advantages in the case study of area IV in the coastal zone. In this case, subsurface terrain modeling in the tidal zone as well as in the sea of the study area was built successfully. It gives a good result because coastline could be used as a breakline to make the model effectively present a terrain surface. However, in any case, a DEM is suited for calculation of slope, aspect, shade relief, and 3D views in this study.

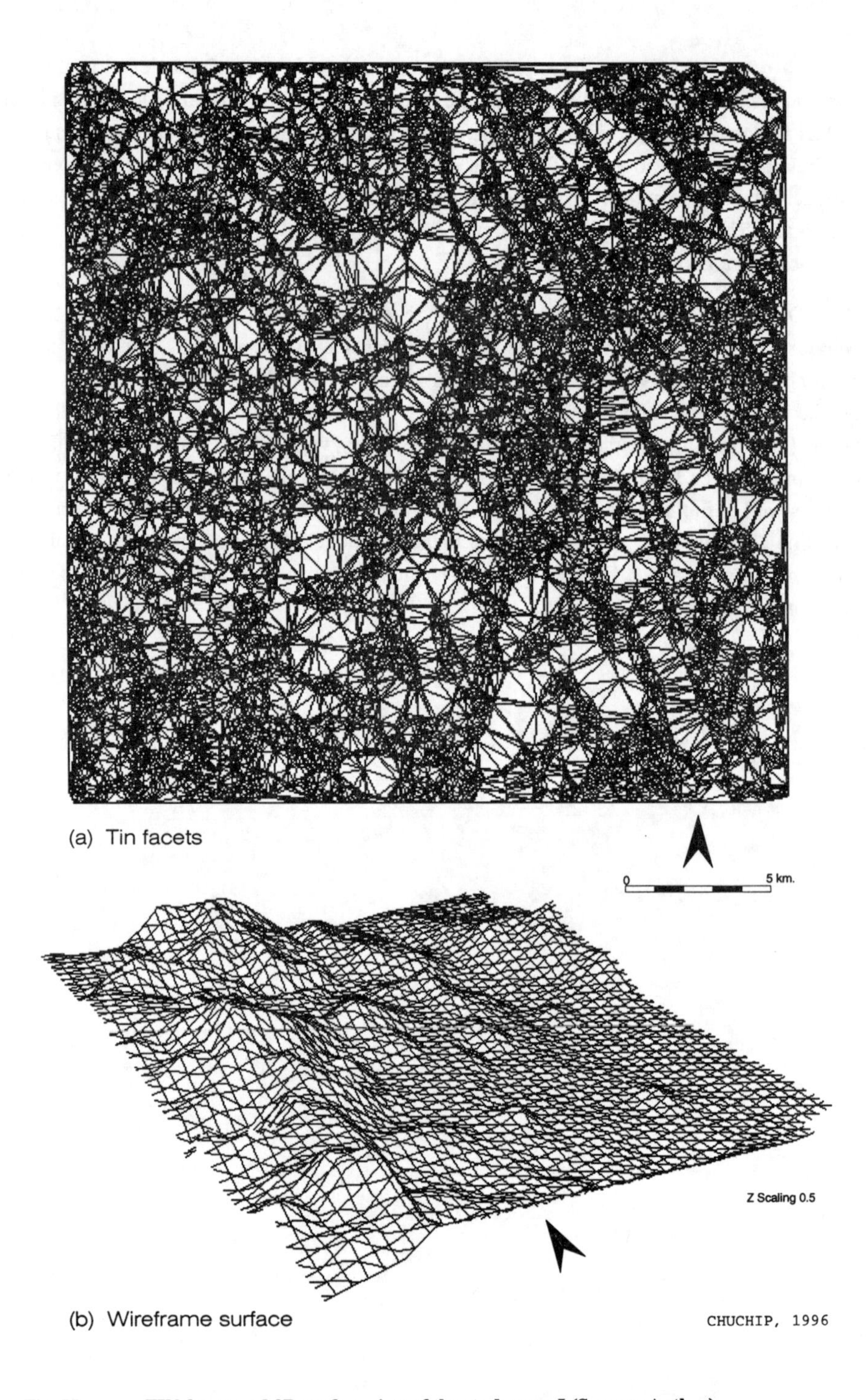

Fig. 55: **TIN facets and 3D surface view of the study area I (Source: Author).**

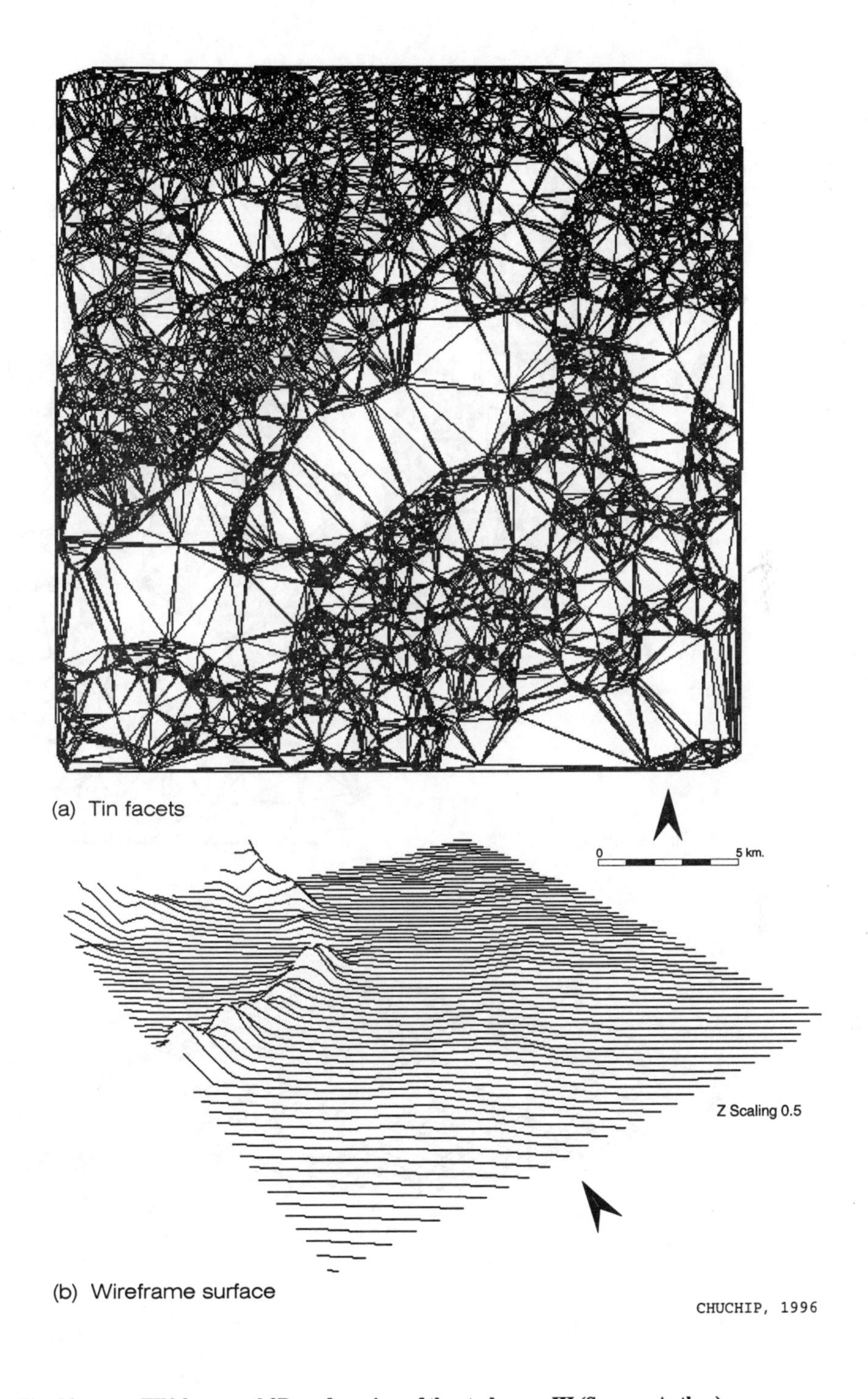

Fig. 56: **TIN facets and 3D surface view of the study area III (Source: Author).**

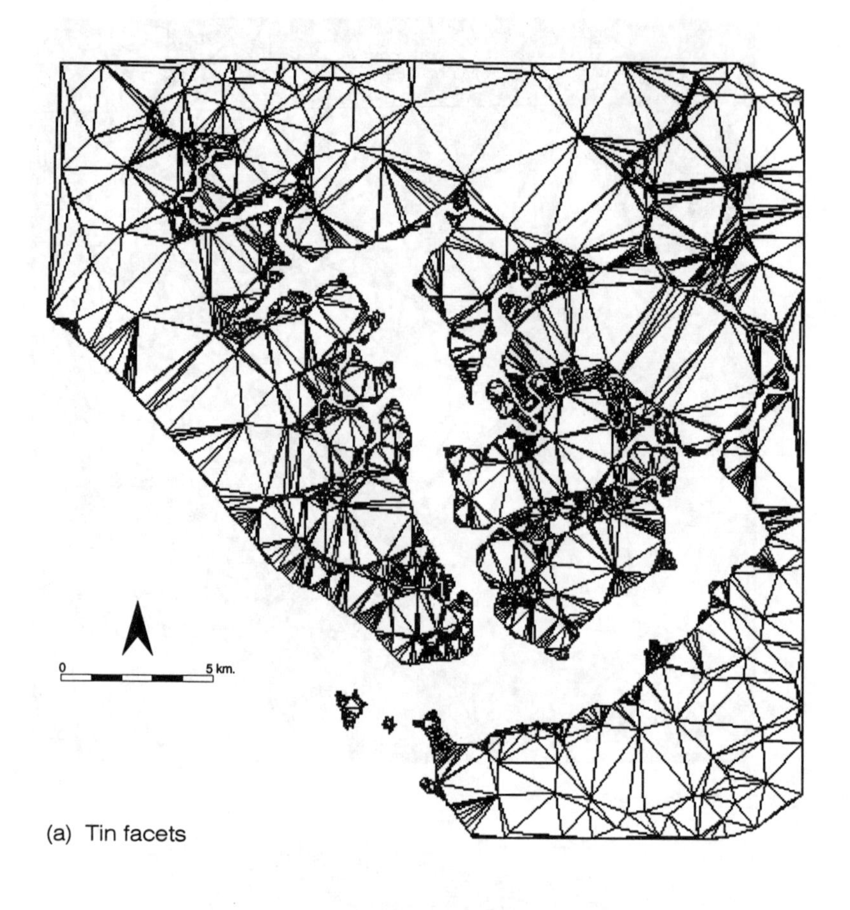

(a) Tin facets

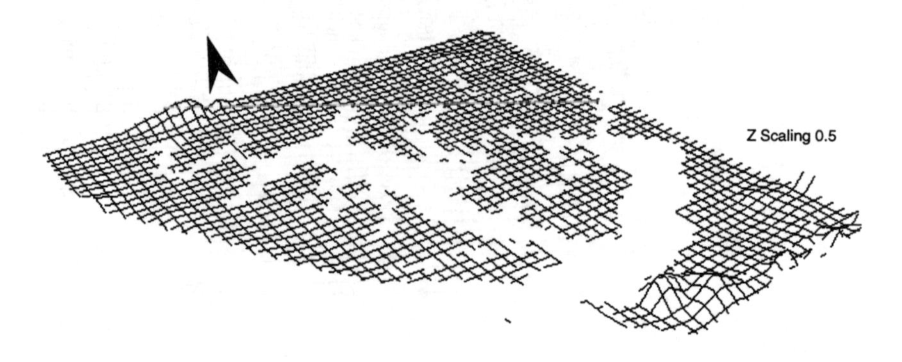

(b) Wireframe surface

CHUCHIP, 1996

Fig. 57: TIN facets and 3D surface view of the study area IV (Source: Author).

As previously mentioned, four areas were selected as the study areas for studying on land use and land cover classification. The physical characteristics of each study area are commonly characteristic for the whole area of the surrounding region. In addition to the 3D view, derived DEM of the study areas were used to produce other topographic data, such as slopes and aspects of the area. With the exception of the Central Plain, derived DEMs were used to interpolate the slope and aspect of the study areas. The slopes ranges and aspects coverage were then calculated for each study areas, as shown in Tab. 27. This information can be also implemented to describe the topography of the study areas.

Tab. 27: Slope ranges and aspects coverage of the study areas (Source: Author).

:Hectare

Slope range	Study area I	Study area III	Study area IV
0-10	36561.562 (58.50%)	60931.625 (97.49%)	61614.187 (98.58%)
10-20	12850.562 (20.56%)	1345.625 (2.15%)	755.562 (1.21%)
20-30	9075.937 (14.52%)	219.500 (0.35%)	93.937 (0.15%)
30-40	2809.437 (4.50%)	3.250 (0.01%)	20.687 (0.03%)
40-50	630.812 (1.01%)	0	14.437 (0.02%)
50-60	217.750 (0.35%)	0	1.187 (0.00%)
60-70	108.437 (0.17%)	0	0
70-80	71.562 (0.11%)	0	0
80-90	49.562 (0.08%)	0	0
>90	124.375 (0.20%)	0	0
Aspect			
East	8569.437 (13.71%)	5683.500 (9.09%)	3786.562 (9.23%)
North East	7414.437 (11.86%)	3949.437 (6.32%)	3902.187 (9.51%)
North	5174.000 (8.28%)	4274.687 (6.84%)	4206.125 (10.26%)
North West	4492.687 (7.19%)	4216.375 (6.75%)	4371.562 (10.66%)
West	5848.250 (9.36%)	2328.875 (3.73%)	4044.687 (9.86%)
South West	6973.187 (11.16%)	3736.125 (5.98%)	4379.312 (10.68%)
South	6348.687 (10.16%)	4703.937 (7.53%)	4147.000 (10.11%)
South East	7499.250 (12.00%)	7407.812 (11.85%)	3789.500 (9.24%)
Flat	10173.750 (16.28%)	26199.250 (41.92%)	8385.812 (20.45%)

Remark: Study area II was not included in this analysis according to its generally flat topography.

Figures 58-60 provide graphical summaries of these data. This information can be used to confirm the description of study areas derived from literature reviews and also can be necessary for further studies relevant to land development planing.

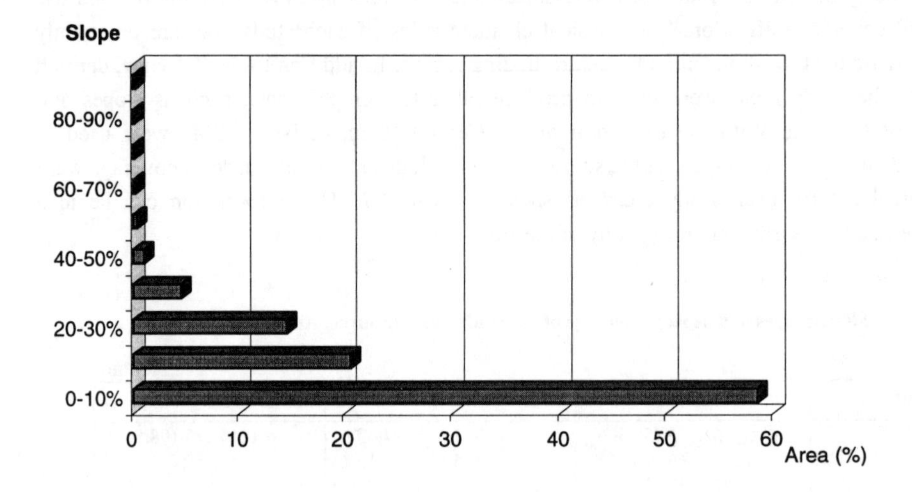

Fig. 58: **Slope ranges in the study area I (Source: Author).**

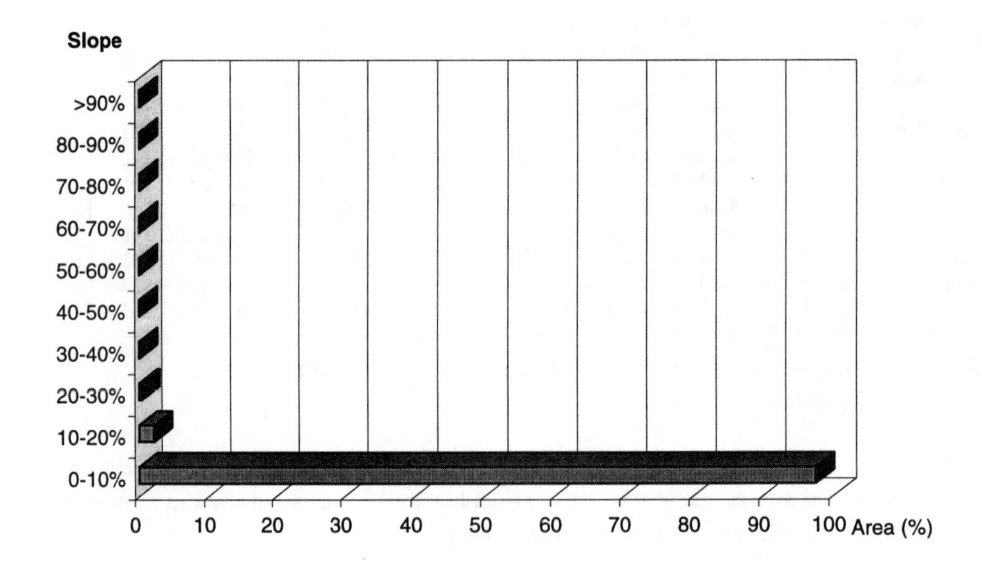

Fig. 59: **Slope ranges in the study area III (Source: Author).**

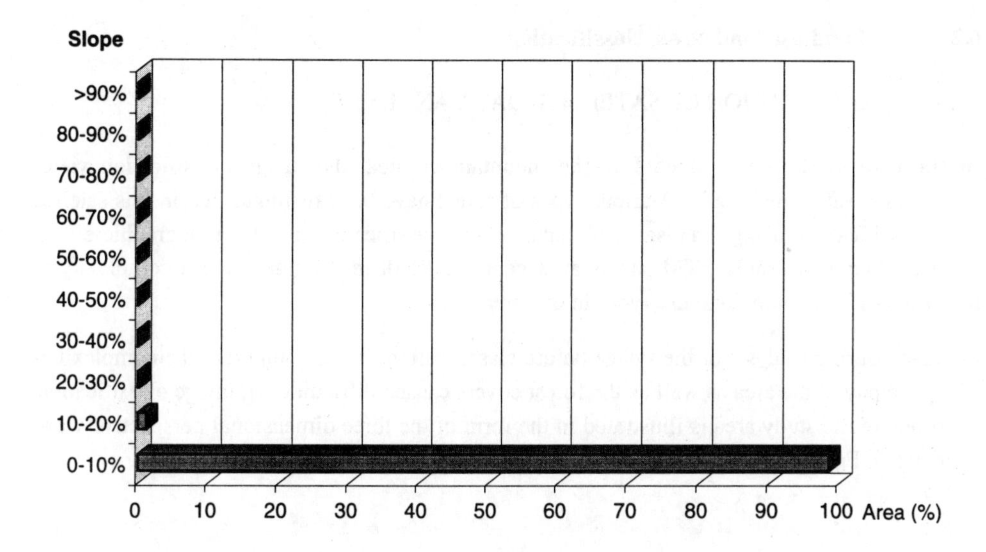

Fig. 60: Slope ranges in the study area IV (Source: Author).

6.2 Land use/land cover classification

6.2.1 EVALUATION OF SATELLITE DATA ANALYSIS

In the case of the study area I in the mountainous area, the image classification yields significantly satisfying results. Various types of forest have been distinguished in this satellite data classification using Landsat TM data. The classification of the topographic-effect-corrected bands of Landsat TM shows an acceptable result in distinguishing a complexity of forest ecology and other land use types in this area.

However, during analysis of the image before classification, it was found that the complexities of topography of the area as well as the forest covers caused difficulties in image classification. A portion of the study area is illustrated in the form of the three dimensional perspective view, as shown in Fig. 61.

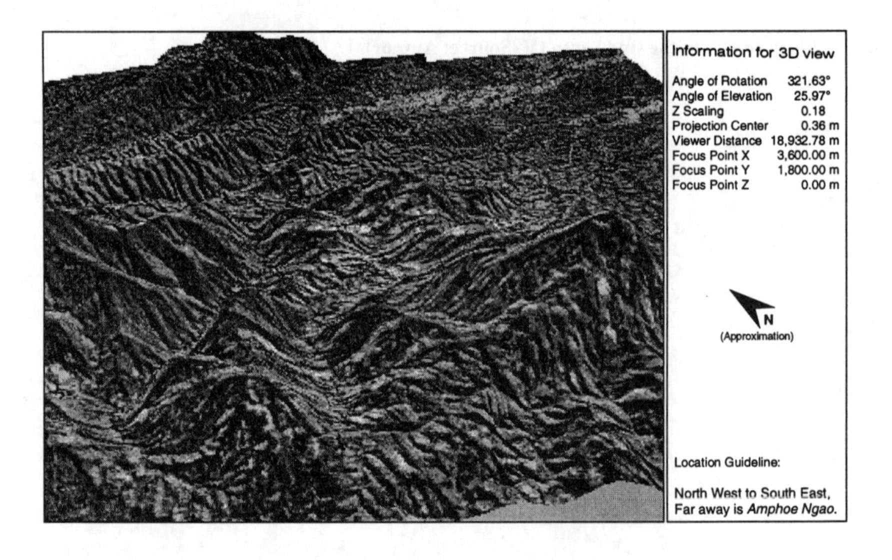

Fig. 61: 3D perspective view of the portion of the study area I (Source: Author).

This 3D perspective view was created from a DEM constructed in this study. Landsat TM band 5, 4, and 3 were draped over the DEM. Topographic effects in the form of shadow aspects can be seen through the entire area. Due to this effect, the same forest types in different aspects have a different illumination.

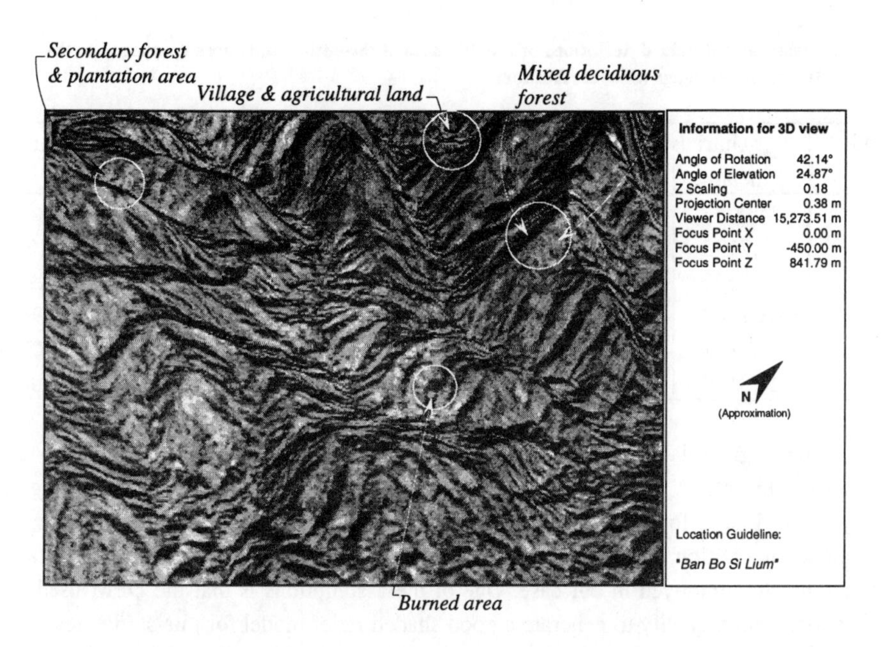

Fig. 62: Land cover types can be easily recognized from the 3D perspective view (Source: Author).

The attempt to alleviate the topographic effect by modifying the digital number (DN) of the original bands of Landsat TM using the shaded relief model created from the DEM was applicable in this study. It was found that, only the first stage of normalization was inadequate to reduce the topographic effect on the image. Thus, the second stage normalization was done in this attempt. Table 28 illustrates the correction coefficients (C_λ) used in the second stage normalization. The mean and variance of both original TM data and the transformed data were compared and presented in Tab. 29.

Tab. 28: Correction coefficients C_λ used in the second stage normalization (Source: Author).

Band	μ_λ	N_λ	N'_λ	S_λ	S'_λ	C_λ
1	66.4307	65.3029	75.8453	67.5659	59.6993	0.1256
2	29.5756	27.9775	32.2553	31.1842	27.2803	0.3928
3	33.5846	31.0000	35.7921	36.1861	31.7324	0.5617
4	64.2369	51.8591	60.2265	76.6961	67.6433	1.4278
5	63.5189	47.1936	55.0205	79.9513	70.4307	1.9059
7	20.1675	14.83152	16.9826	25.5384	22.1664	2.0367

Table 29 indicates that the overall spectral reflectance properties of the original TM data are preserved with the normalization while the topographic effect was partially removed. Figures 63 and 64 show the band combination of the original TM data in comparison to the resulting image after applying two-stages normalization. It can be seen that topographic effect is

Tab. 29: Means and standard deviations of satellite data of the entire study area compared to the transfromed image (Source: Author).

Band	Untransformed		1st stage normalization		2nd stage normalization	
	mean	SD	mean	SD	mean	SD
1	74.0057	6.8229	73.6165	9.2645	73.6839	6.9384
2	33.7185	3.9267	33.3425	4.7260	33.3838	4.0713
3	41.5539	7.9213	41.1568	8.5283	41.1901	8.1119
4	63.1767	12.2930	62.5079	11.9249	62.3581	12.6194
5	77.2883	20.8564	76.5227	20.7070	76.0987	22.2800
7	28.3179	10.9849	27.8559	11.0524	27.7013	11.5908

apparent in the original data, especially in the southerly slopes. After applying the two-stages normalization, this effect is reduced. Actually, this showed the success in improving the capability of an image for best use with the visual interpretation or for the unsupervised classification. In addition, it must be noted that the topographic effect from the TM data has been not completely removed in our case. One of the assumptions is that the DEM used may have not good enough quality to generate a good shaded relief model for this study area where the topography is very complex. The other reason is about the affectation of the rock outcrops occurring in this area. As stated before, the deciduous forest is the main land cover occurring in this region. On the top of the mountains some parts of this area are mostly covered with outcrops of limestone. Landsat image used was acquired in the dry season in which the deciduous forests were partially defoliated. The differential illumination of southerly and northerly surfaces varied greatly. As a result, unsupervised classification with this transformed image did not result a satisfying classified image. Northerly facing aspects still appear shadow-effect categories throughout the classified image. Thus, the interactively supervised classification was then used to classify the image.

Categorizing the secondary forest into different regenerative stages was not success, as forests of different ages had quite similar spectral characteristics. This is because only the single date Landsat TM imagery was used for the classification. This was also the case for with categorizing the forest plantation. Furthermore, the patterns of forest regrowth and young forest plantation were mixed together. Both planted trees and regenerating trees were found in the same area. This is because weeding has not regularly been done in the plantation areas. The planted trees were mostly covered with herbaceous plants. Thus, this can cause a problem in the image classification. The forest plantation area would be miss-classified as the secondary forest or an old clearing area.

Mature teak plantation area was virtually impossible to differentiate from neighboring natural forest which is normally a deciduous forest with teaks as a dominant species. Dry evergreen forest was possible to discerned with difficulty, but it is hard to test the reliability of this interpretation without checking in the field or determining ancillary data.

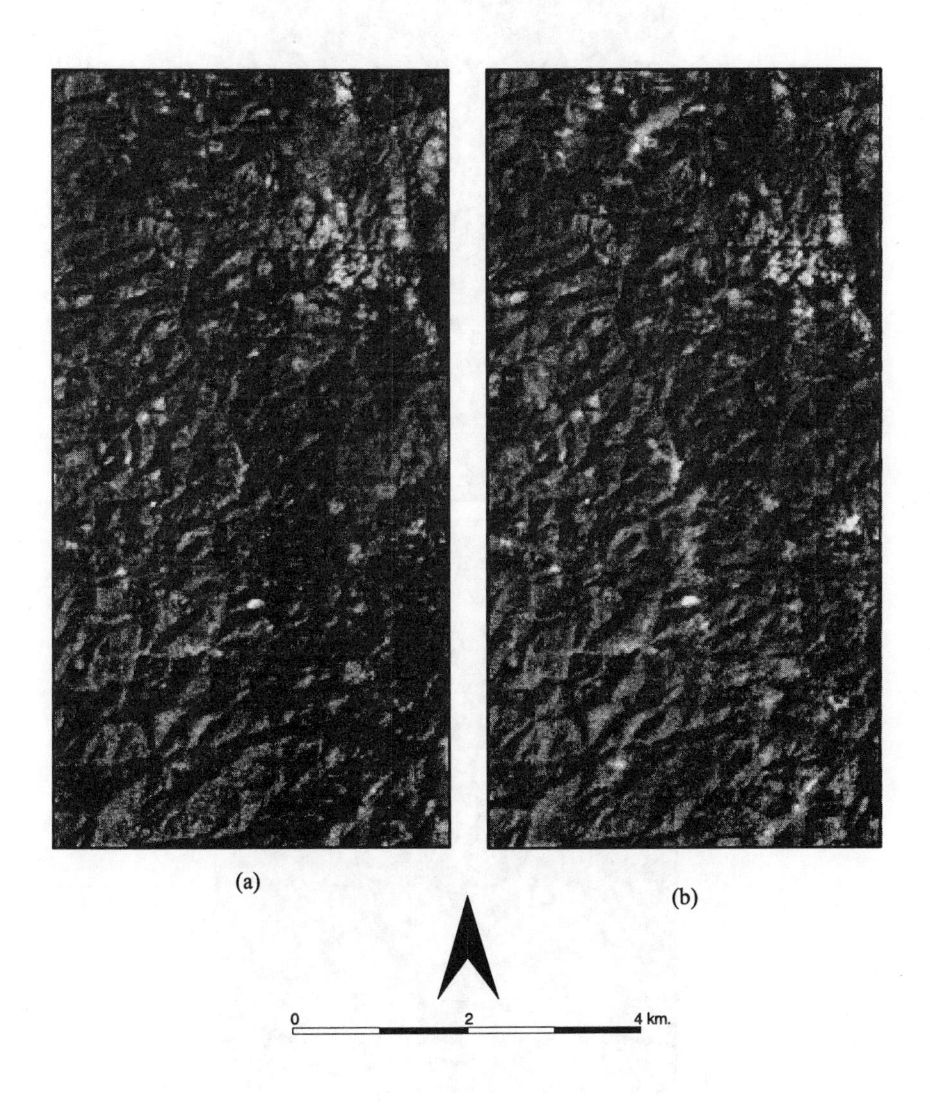

Fig. 63: **Band combination 4, 3, and 2 of the portion image of the study area I, (a) Original TM band (b) Result of the two-stages normalization using a DEM (Source: Author).**

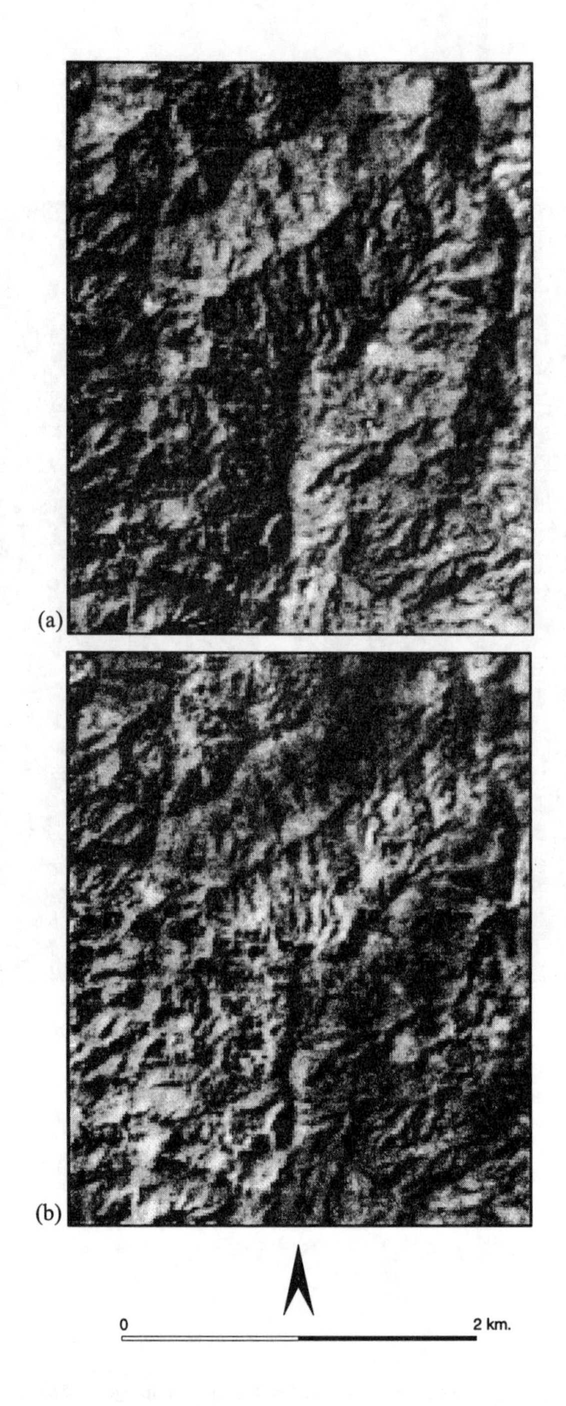

Fig. 64: **Band combination 5, 4, and 3 of the portion image of the study area I, (a) Original TM band (b) Result of the two-stages normalization using a DEM (Source: Author).**

Separating deciduous forest category into mixed deciduous forest and dry dipterocarp forest was possible. Furthermore, subclasses of these categories could be established. The status of these forests are different according to the clown closure density and the tree species composition. This also includes the disturbance status, such as forest fire damages and forest logging, which could be distinguished. Many shifting cultivated areas were found in the forest areas. These man activities can easily cause to the status change of neighboring forest areas.

Paddy fields and field crops both in the plains and the highlands are the main types of agricultural categories of this classification. They were discriminated without any difficulties. The developed crop calendar (Fig. 41) is significantly helpful. It improves the capability of the image classification during the study.

The urban category in terms of built-up areas has not been found in the image. The discrimination of small villages was impossible. It can be noted that the neighboring pixels of village pixels made the village boundaries seen in the classified image, but villages pixels are a class similar to the paddy field class. Thus, post classification could be done to derive this category.

Water was also possible to be classified. Only small reservoirs were found in this area. They can be distinguished without difficulties. However, the signature of water class caused also miss-classified pixels in areas with deep shadow slopes. This needed to be corrected with post-classification.

In conclusion, the level III of the land use and land cover classification shown in Fig. 43 were successfully derived using Landsat TM data with their transformed bands and the interactive supervised classification algorithm. It also seems that the best result was derived after the use of TM data had been made through a multilevel approach combining original TM data, enhancement techniques, DEM, and other available ancillary data.

Table 30 summarizes the land use and land cover categories distinguishable on Landsat TM data by computer-aided analysis. The categorization is based on the land use and land cover classification system, developed in the early phase of the study as shown in Fig. 43.

Tab. 30: Land use/land cover categories distinguishable on Landsat TM data by computer-aided analysis method for the study area I.

Level I	Level II	Level III	Level IV
Forest land	Evergreen forest	Dry evergreen	
	Deciduous forest	Mixed deciduous	Dense crown closure
			Disturbed (most from logging)
			Dominant with bamboo (heavily disturbed)
			Old clearing and naturally regenerating forest
		Dry dipterocarp	Undisturbed Disturbed (from clearing and forest fire)

Tab. 30: (continued)

Level I	Level II	Level III	Level IV
Agricultural-land	Paddy field	Harvested rice field appearing field crops	
		Dry, harvested paddy field	
	Field crop	Upland crop	
Urban area	Village*		
Water	Reservoir		

* after applying post-classification

Figure 65 shows the proportion of the main land use and land cover categories derived from the classification. It was indicated that mixed deciduous forests occupied most areas in this study area. It can be noticed that this forest type could have been severely disturbed in the past. Stumps of big trees were widely found in the mixed deciduous forest with bamboo. The proportion of forests shows that the condition of forest in this area is endangered, in particular the Mixed deciduous forests.

Using GIS overlay, the relationship between soil types of this area and the land use and land cover classes was graphically summarized in Fig. 66. The land cover types in this area can be identified in combination with certain soil types. This relationship could be essential for the further development in this area. The site suitability for reforestation by considering the soil factor is an example of using this relationship. In addition, it can be noticed that the 'old clearing' category was found in the Slope Complex (SC)and Li soil series (Li). These soil types are sensitive for erosion. The resulting DEM was used to calculate the slope ranges and aspect of the study area. These resulting terrain data have been overlaid with the classified image. The relationship between the terrain data and land use and land cover categories are summarized in Fig. 67 and Fig. 68.

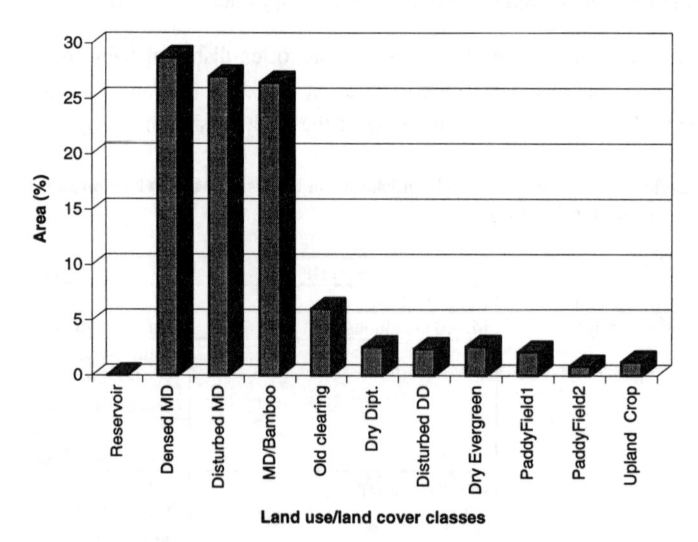

Fig. 65: Proportion of land use and land cover types in the study area I (Source: Author).

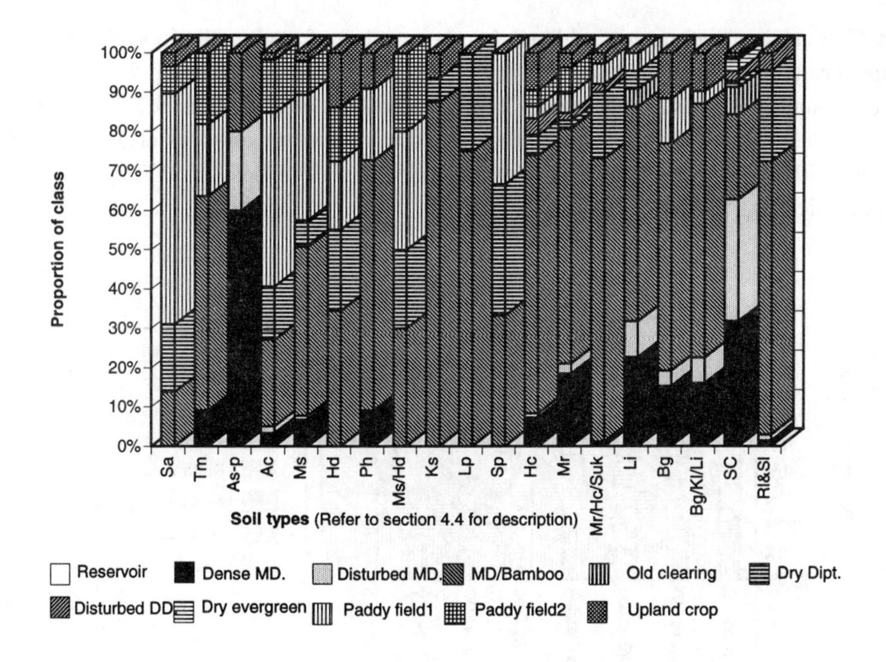

Fig. 66: **Relationship between land use/land cover types and soil types in the study area I (Source: Author).**

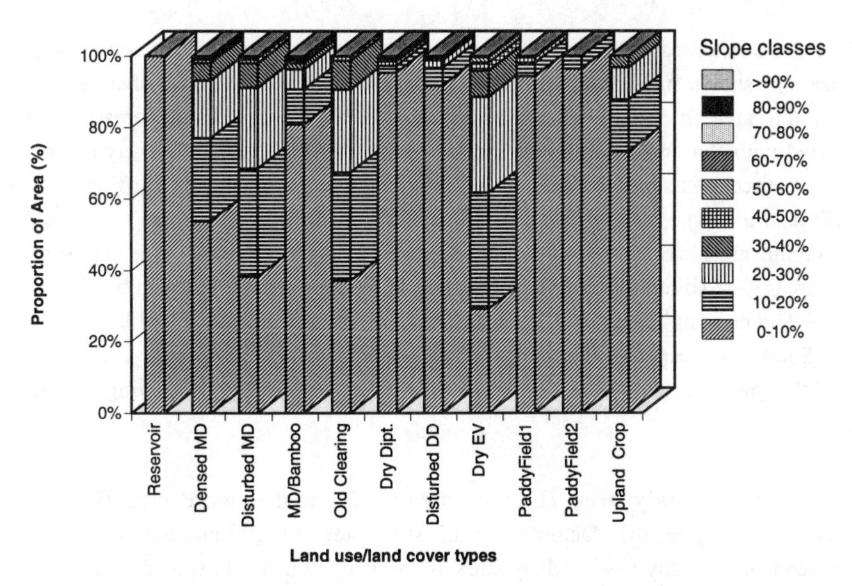

Fig. 67: **Relationship between land use/land cover types and slope classes of the study area I (Source: Author).**

Figure 67 illustrates that mixed deciduous forests have been found in all slope classes, while dry dipterocarp forests were mostly found in the slope range 0-10%. Dry evergreen forests have been also found in various slope ranges. These conclusions are corresponded with the results from the field survey. It should be noted that the upland crops category was also found in the high area with a slope over 20 %. These cropping areas are a kind of shifting cultivation practices which may cause soil erosion.

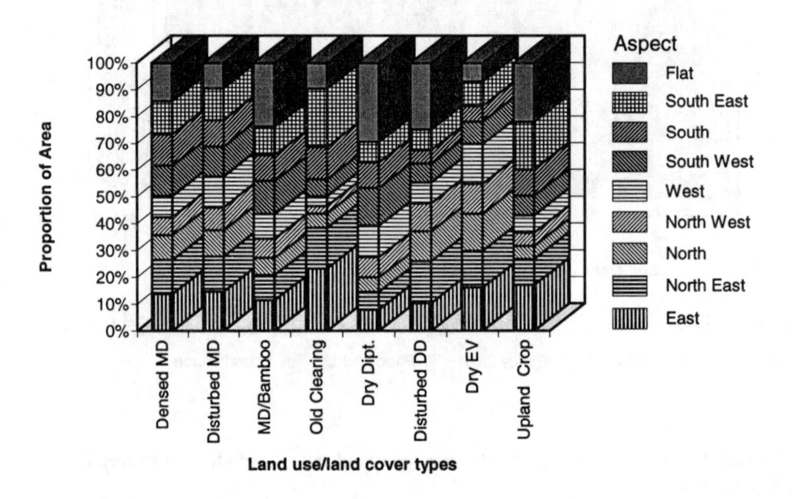

Fig. 68: Relationship between forest cover types and the aspect of the study area I (Source: Author).

In addition, the relationship between forest types and the aspects of this area can be also summarized, as shown in Fig. 68. Normally, the aspect of area could also characterize forest types. For this area, it can be seen that all types of mixed deciduous forests are mostly characterized with flat and East aspect. Dry dipterocarp forests area are mostly characterized with the flat aspect. As stated before, upland cropping areas were mostly shifting cultivation. Thus, this land use type was included in this evaluation. In can be noticed that most of the upland cropping areas were found in the flat, South East, and East aspect. This can be assumed that these areas were suited for cropping according to the light factor. This can be also linked to the case of 'Old clearing' category. This land use type was also mostly found in the areas with East and South East aspects, the same as the 'Upland cropping' category. This could be concluded that most of the old clearing areas were used for cropping or shifting cultivation in the past.

In the case of the study area II in the Central Plain, the image classification yields significantly satisfying results. Determining this study area, it was found that nearly the whole area is covered with paddy fields. Many streams and irrigation canals spread through the area. From the crop calendar it is evident that rice is grown all year round. The growth of rice in paddy fields is commonly found in different status varying from the site preparation stage to the harvesting stage according to the water supply from irrigation systems and the aims of the farmers. The condition of growth commonly differs in this area. Owing to these factors, the image classification was not so easy as expected.

It was also found that some portions of paddy fields contained pixels having reflectance values similar to the pixels of water areas. Furthermore, other part of paddy fields are heterogeneous. Hence, unsupervised classification was performed first. As expected, pixels in paddy field areas were grouped into many categories. It can be noticed in this stage that there are some factors affecting these areas. To make clear, a DEM constructed in the first phase of the study was used to interpolate the contours with the interval of 0.2 meter (as shown in Fig. 69).

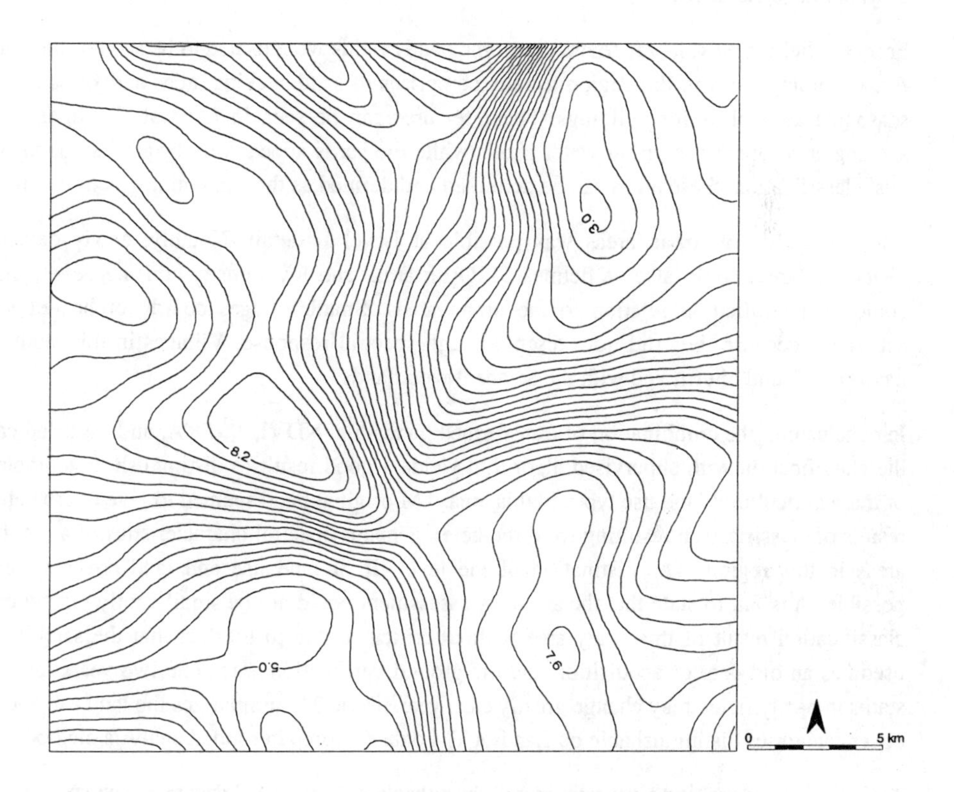

Fig. 69: Contour lines interpolated from a DEM of the study area II (Source: Author).

Resulting contour lines were overlaid on the TM image. It shows that the area is not so perfectly flat. Thus, it can be assumed that paddy fields may be flooded in some areas while the other parts are drier. This was confirmed after the field work had been done. The people in this area pointed out that not the whole area is flat. The local farmers usually select the optimal rice species to grow in these leveling fields to prevent the rice production damage in the raining season. The farmers coarsely differentiate these rice species into two types, locally called 'small rice' and 'tall rice'. Furthermore, in the northwest of the area, farmers usually select sugar cane for planting in this zone instead of rice. This is because this part of area is relatively higher than the other parts. Water is sometimes lacking for growing rice. Some parts of this area are harvested paddy fields with bare soil. Some portions of these region are ploughed for cropping. They are mostly used for sugarcanes. It can be noticed that some areas are ploughed while some are bare soils. These soil conditions can be distinguished from the image. These areas were classified into harvested paddy field categories owing to their similar reflectance values. Furthermore, it is interesting that some parts of this area that are actually sugar cane fields were covered with water. From field survey, it was clear that these areas were harvested sugarcane fields that were already harvested. Farmers have filled the water into these areas in order to let the sugarcane stems sprout again. These caused these areas being classified as water class in image classification.

From the field survey, it was found that some paddy fields were turned to be sand mining areas. Active mining areas in this region usually comprises of sand sucking areas and yards for sand seasoning as well as for buildings. These features can be seen in Landsat TM image. Sand sucking areas appear as turbid water pixels while the yards appear very bright bare land. With post classification, these areas can be delimited and defined as the sand mining activity class.

The delineation of urban areas was possible at Level II detail. Notably, newly developed residential areas could also be delineated. Open areas and low-density housing developments could be identified in relation to the urban areas. Small villages could not be delineated accurately because they did not present enough spectral response. Villages in this study area have been found intermixed with the adjacent woodlands.

In conclusion, the combination of transformed bands, i.e. NDVI, 1st PCA, and Tasseled cap in the classification with supervised algorithm yields a good result to distinguish the complexity of these agricultural land use types in this area. TM original band seemed to give overwhelming results of classification resulting from the heterogeneous spectral characteristics of agricultural areas in this region. The distinction of the level III of land use and land cover system is possible. It is fair to state that the accuracy assessment could not be significantly judged by the classification result of this study area without criticism, due to the fact that the satellite data used has an old date of acquisition and the ground verification was done two years later. The status in paddy fields may change greatly each year. Table 31 summarizes the land use and land cover categories distinguishable on Landsat TM data by computer-aided analysis method.

Tab. 31: Land use/land cover categories distinguishable on Landsat TM data by computer-aided analysis method for the study area II.

Level I	Level II	Level III	Level IV
Agricultural land	Paddy field	Rice field in growing stage	In high water stage
			Young stage with wet soil
			Mature stage with wet soil
			Mature stage with dry soil
		Harvested rice field (Bare soil)	
	Field crop Trees/home gardens	Sugarcane	

Tab. 31: (continued)

Urban areas	Built-up areas		
	Open-urban	(Mixed with infrastructures, i.e. roads)	
Water	Clear water	River (Spatial recognition) Canal (Spatial recognition)	
	Turbid water	(It was found from field survey that this class is the area of sand mining)	

The proportion of land use and land cover types occurring in this area can be summarized as shown in Fig. 70. The figure illustrates that this study area is mostly covered with paddy fields in different stages. Paddy fields in the mature stage with relatively dry soil were mostly found. It can be noticed that varying stages of the paddy fields in this area could cause a problem in verifying the result of satellite image classification. The status of these land use types could change at any time. Ground verification must be done in a proper time and as early as possible.

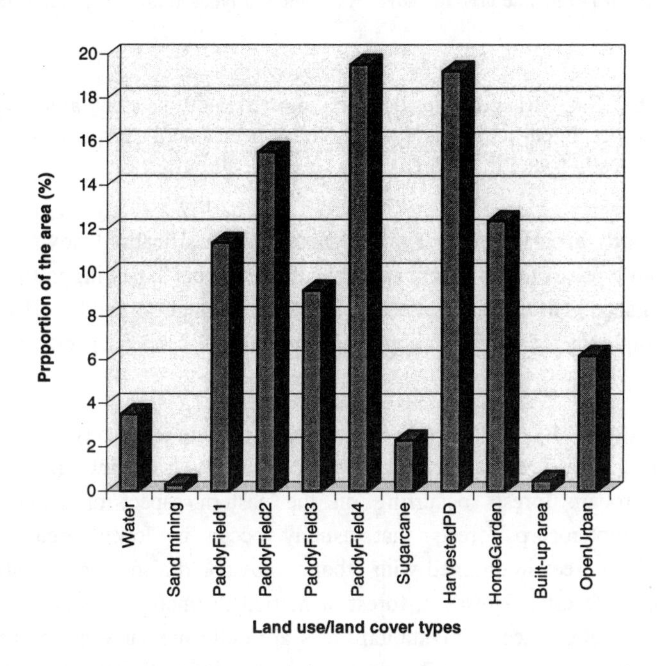

Fig. 70: Proportion of land use and land cover types in the study area II (Source: Author).

Since this study area is an important agricultural area, the categorization of agricultural land uses was also identified in the level III of the classification system. Furthermore, the relationship between the land use/land cover types and soil series was studied, using the GIS overlay method. This could be beneficial for further investigate in this study area. Figure 71 illustrates this relationship.

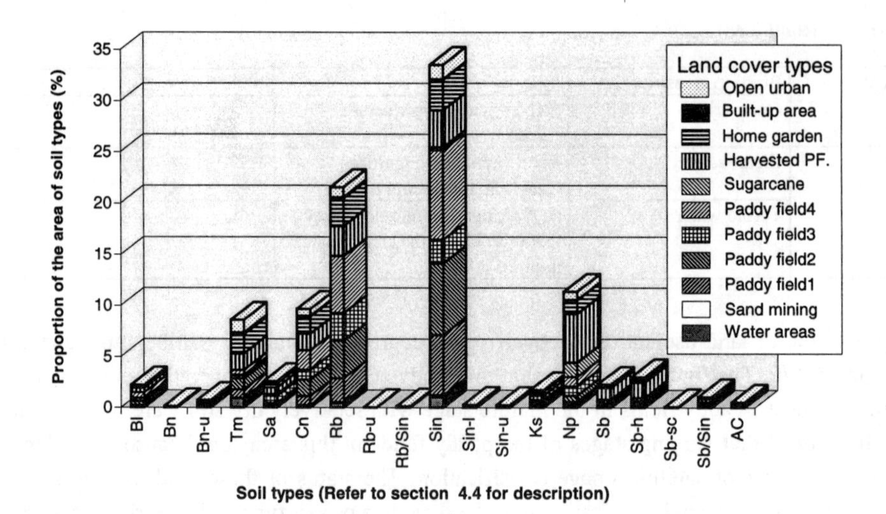

Fig. 71: Relationship between land use/land cover types and soil types in the study area II (Source: Author).

It can be concluded that, Singburi soil series occupied most area of this study area. Many agricultural land use types, in particular paddy fields, have been found over this soil series, while sugarcane was mostly found in the areas with Nakorn Pathom series.

In the case of the study area III in the Korat plateau, the classification shows significantly good results. All main types of forest covers and agricultural covers appearing in this area were successfully distinguished. TM original reflective bands are enough to yield a good result for distinguishing a complexity of forest ecology and agricultural land uses occurring in this plateau.

Dry evergreen forest and dry dipterocarp forest are the main forest types that can be differentiated in the Landsat TM. Categorizing the dry deciduous forest into different strata is possible. High dipterocarp forests occurring on the eastern aspect of mountains were distinguished. Low dipterocarp forests that usually occur in lower areas were also distinguished. Deciduous trees intermixed with other scrubs could also be differentiated from the low dry dipterocarp forests. However, forest plantation distinction was not successful in this classification. The reflectance of plantation areas are quite the same as the neighboring forests. Brushes and shrubs have also reflectance values similar to the evergreen forest. The complete separation these land cover types was not possible with the normal classification scheme. Miss-classified pixels have been found. Fortunately, these forest types commonly occur in different locations in this area. Brushes and shrubs were found only in plains near streams while the dry evergreen forest was found only in the mountainous areas.

Fig. 72: Topographic effects appearing in the Landsat image can be seen from a 3D perspective view (Source: Author).

Due to the topographic effects, land cover types on the shadow slopes had to be assigned to a category different from the same types occurring in the another side. Some isolated pixels occupying a water class have been found in the high areas. These pixels are considerably miss-classified. However, these pixels were successfully eliminated by applying filtering in the post-classification. Figure 72 illustrates the topographic effects occurring in some parts of this study area. This image is the 3D perspective view using a DEM superimposed with the Landsat TM band 5, 4, and 3. On the image pixels in the shadow areas and pixels in the water bodies have rather similar digital numbers.

Paddy fields and field crops are the main types of agricultural covers found in this area. They were discriminated without any difficulties. Furthermore, the paddy field class can be categorized into a number of subclasses based on geomorphology and landforms. This was succeeded because the paddy fields were in the harvested stage, soil parent materials and landforms could affect the overall reflectance of these areas. Sugarcane is the main field crop found at the time of satellite data acquisition. Distinction of sugarcane fields yields good results. Figure 73(a) illustrates a portion of Landsat data band 5, 4, and 3 that were draped on the 3D perspective view of the DEM. It shows that the sugarcane fields can be easily seen both from their spectral and spatial characteristics. The sugarcane fields appear different from other land use/land cover types, e.g. low dry dipterocarp forests, bare soils, on the image. The developed crop calendar (Fig. 42) is significantly helpful. It improves the capability of the image classification during the study.

Urban settlement categories in terms of built-up areas and villages can be easily delineated. The discrimination of a small village was also successful. Villages in this area are commonly surrounded by trees or home gardens. This can also be seen from Fig. 72. These neighboring pixels of vegetation reveal the spatial characteristic of the villages in the image. Thus, without post classification this category can be obtained in this area. Transportation areas such as roads can also be identified, but only by association with other neighboring features.

Idle land such as rock-outcrops and bare soils were also possible to be discriminated. Water bodies such as reservoirs, ponds, and streams were distinguished without any difficulties. However, there are a number of pixels classified as water class appearing in some areas in deep shadow slopes. These miss-classified pixels are sparse and isolated. Thus, filtering was done to eliminate these pixels. The appearance of the reservoirs and streams can be seen easily as shown in Fig. 73(b). Shrubs commonly found surrounding reservoirs and streams in this area were distinguished from the image. It was found that they were miss-classified as evergreen forest. This was because the reflectance values of these land covers are quite similar in this case. Thus, post classification is necessary to correct this miss-classification.

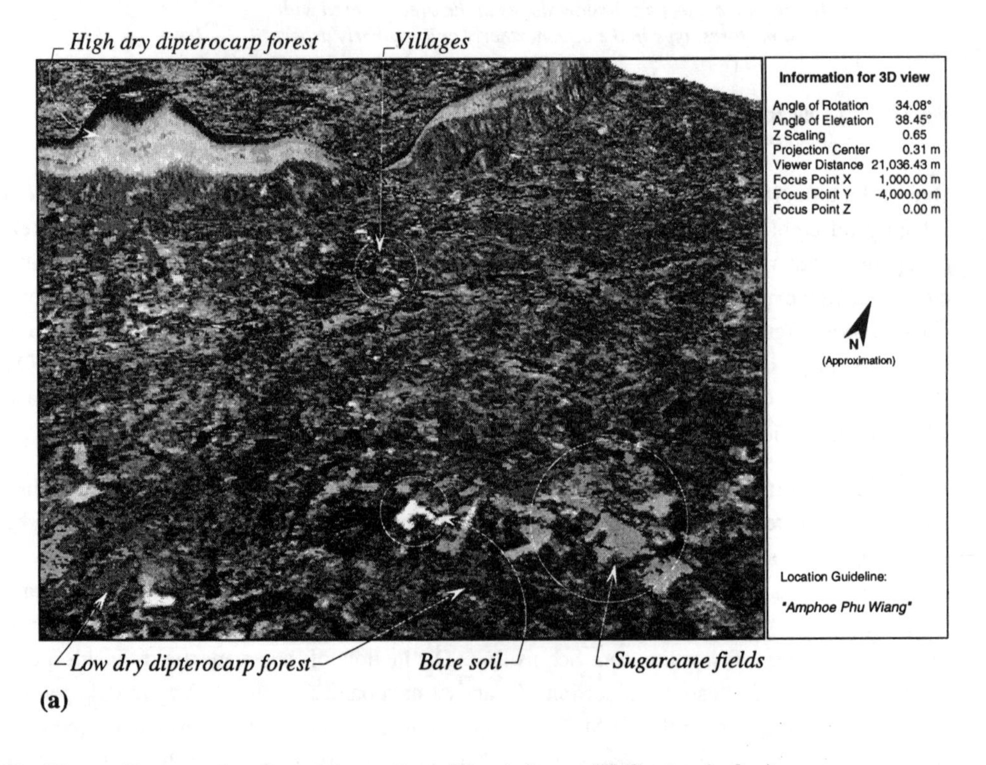

(a)

Fig. 73: 3D perspective views of the portions of the study area III (Source: Author).

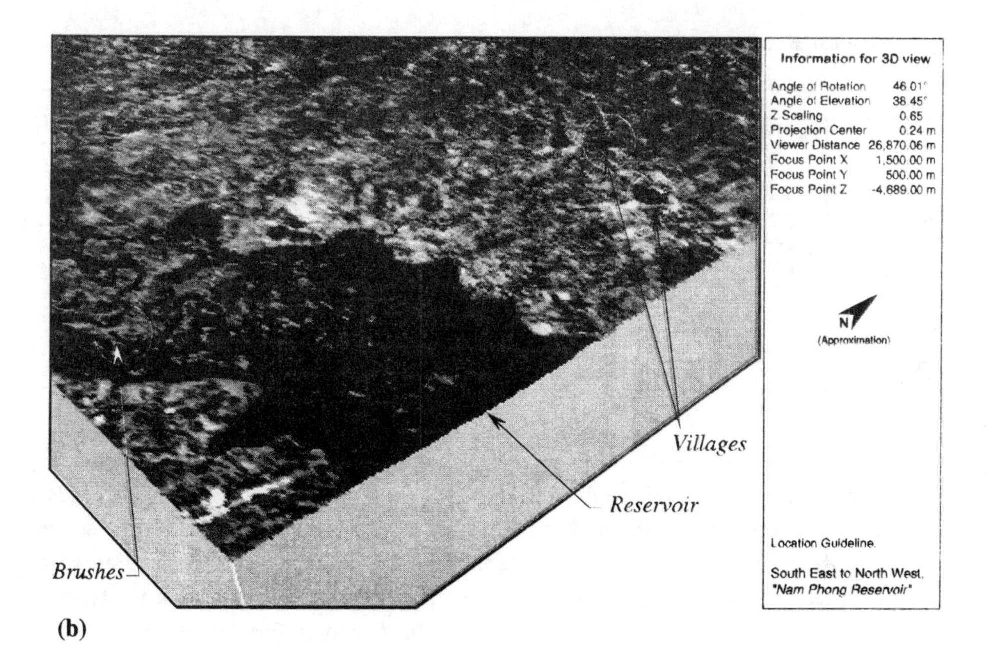

Fig. 73: **(continued)**

In conclusion, the level III of the land use and land cover classification shown in Fig. 43 were successfully derived using the original bands of the Landsat TM data with the supervised classification algorithm. The transformed bands such as NDVI, Tasseled cap, were not necessary to be implemented in this study area.

Table 32 summarizes the land use and land cover categories distinguishable on Landsat TM data by computer-aided analysis method. The categorization is based on the land use and land cover classification system developed in the early phase of the study as shown in Fig. 43.

Tab. 32: **Land use/land cover categories distinguishable on Landsat TM data by computer-aided analysis method for the study area III.**

Level I	Level II	Level III	Level IV
Forest land	Evergreen forest	Dry evergreen	
	Deciduous forest	Dry dipterocarp	High dry dipterocarp
			Low dry dipterocarp
			Deciduous trees mixed with scrubs
Range land	Shrubs, brushes		
Agricultural-land	Paddy fields	Harvested	In low floodplain
			In low alluvial plain
			In undulating terrain
			In erosional surface

-continue-

Tab. 32: (continued)

Level I	Level II	Level III	Level IV
	Field crops	Sugarcane	
		Other cashcrops	
		Upland crops	
Idle land	Bare soil		
	Rock-outcrops		
Urban land	Built-up		
	Villages		
Water	Streams	(Can be identified to: deep, shallow, turbid)	
	Reservoirs		
	Ponds		

The proportion of land use and land cover types occurring in this area are summarized in Fig. 74. The figure indicated that this study area is mostly covered with harvested paddy fields in alluvial plains. Paddy fields in the middle undulating terrain were also found in a wide area. Since the image used was taken in dry season, most paddy field areas in this study area were mainly characterized with the degree of spectral characteristics of soil. This land use type was successfully categorized in level III by differentiating the geomorphology of the area.

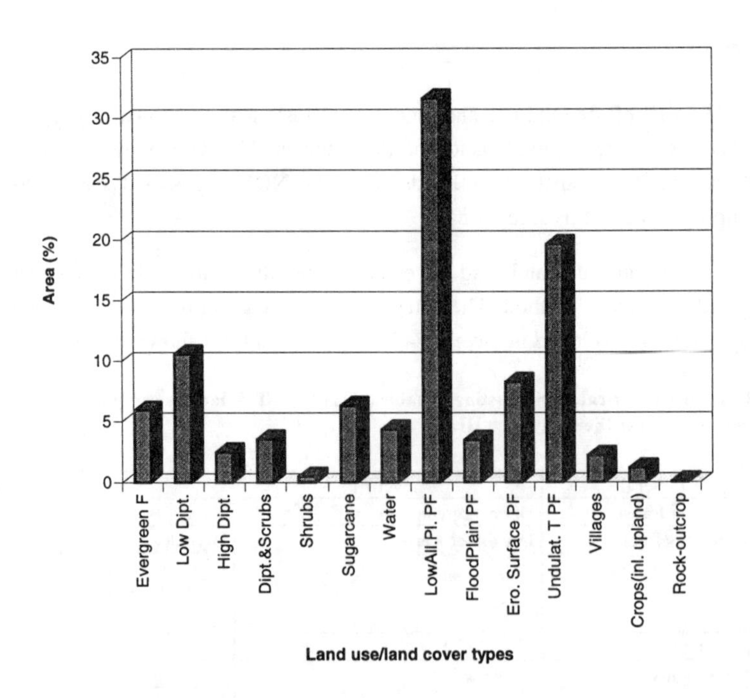

Fig. 74: Proportion of land use and land cover types in the study area III (Source: Author).

To study the relationship between land use/land cover categories and the geology of the study area, a simple matrix analysis was performed. Resulting data were graphically illustrated in Fig. 75. This analysis also confirms the results of the land use and land cover classification in this area. It can be noticed that some land cover types have been influenced by rocks types. For example evergreen forest, high dipterocarp forest, and Rock outcrops occurred mostly over the Phra Wihan Formation (Jpw) and Sao Khua Formation (Jsk). These rock types were commonly found in the mountainous areas. From field survey, it was also found that these land cover types occurred mostly on the high mountain areas. Furthermore, it can be seen that low dipterocarp forest was mostly identical with the Phu Kradung Formation and Nam Pong Formation the same as paddy fields. Based on this relationship it can be noticed that many paddy field areas may cover with the dipterocarp forests in the past. From field surveys, it can be seen that many dipterocarp trees can be found everywhere among the paddy fields. Most of paddy fields and crop areas cover nearly the same rock types. These are the Phu Kradung Formation, Nam Pong Formation and alluvial deposits. Further matrix analysis was then carried out using the soil types of this area. Figure 76 illustrates the result of the analysis. It can be noticed that the Roi Et series mostly found in this area were used for paddy fields. The Nam Pong series is also covered with paddy fields and cropping areas. It can be concluded that the categorization of harvested paddy fields in this region at the level III could not be based on a single factor, such as soil type or geology. Categorizing this class may have to consider the spectral response of these areas. The geological background and soil conditions of the area could be considered.

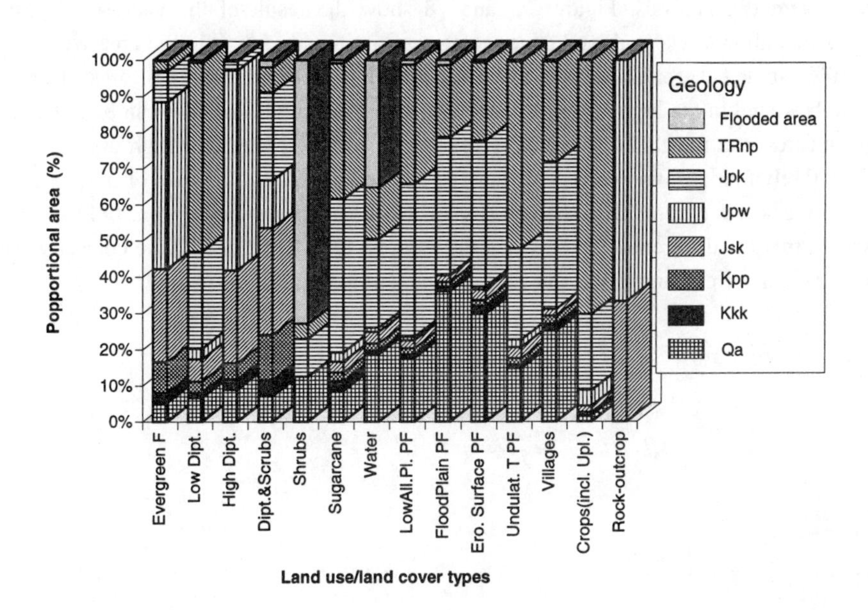

Fig. 75: Relationship between land use/land cover types and rock types in the study area III (Source: Author).

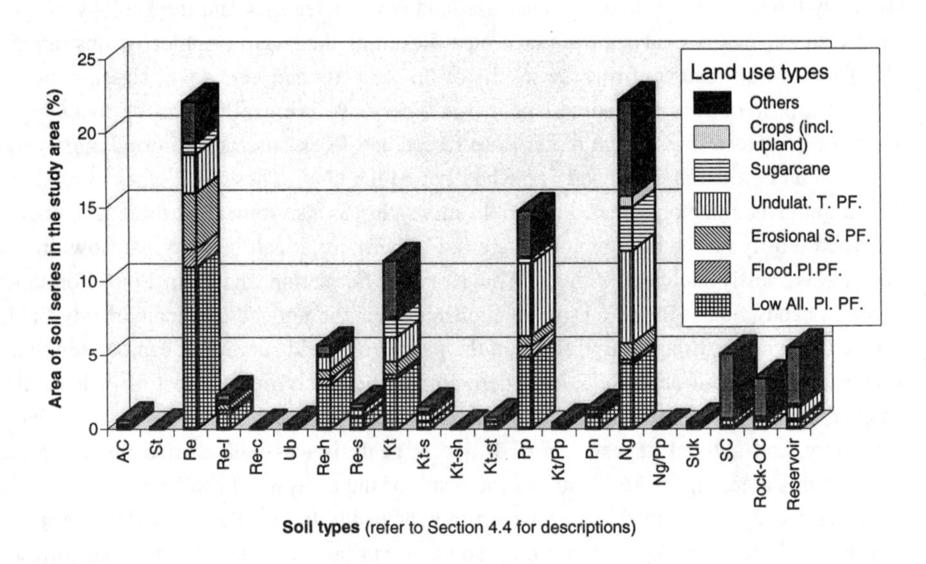

Fig. 76: Relationship between the land use types and soil types of the study area III (Source: Author).

The relationship between terrain data and the land use/land cover categories was also carried out using a matrix analysis. Figures 77 and 78 show the result of the analysis. Figure 77 illustrates that all land use and land cover categories are mostly found in the area with a slope below 10%. Some categories, such as evergreen forest, high dry dipterocarp forest, Rock out-crop, can be found in the high area with a slope over 10%. From Fig. 78, it can be noticed that evergreen forest can mostly be found in the North West aspect, while the high dry dipterocarp forest are rarely found in this zone. This may be a shadow effect in the case of other area. The North West aspect is commonly a shadow side of the mountainous areas. From field surveys, it was shown that the differentiation between these forest types have been done correctly. These land cover types were characterized by geology and soil factors.

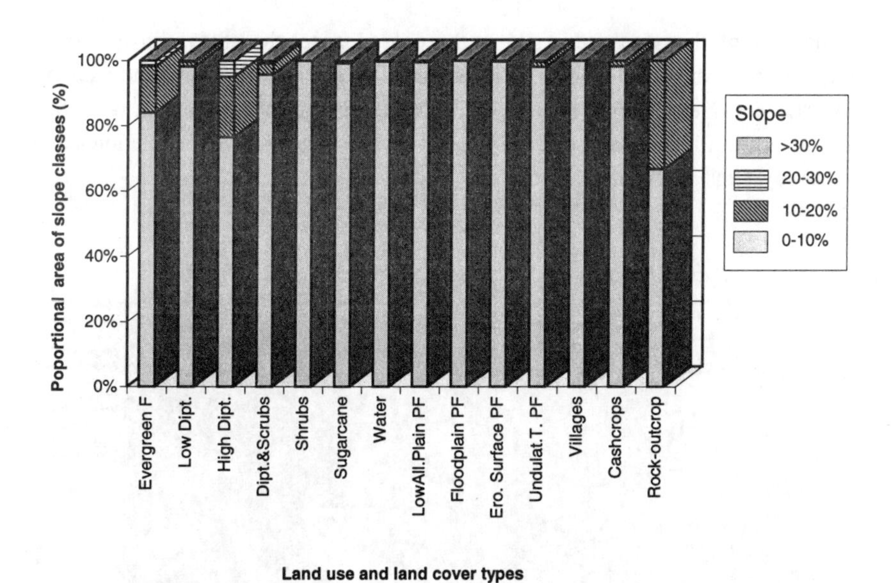

Fig. 77: **Relationship between land use/land cover types and slope ranges in the study area III (Source: Author).**

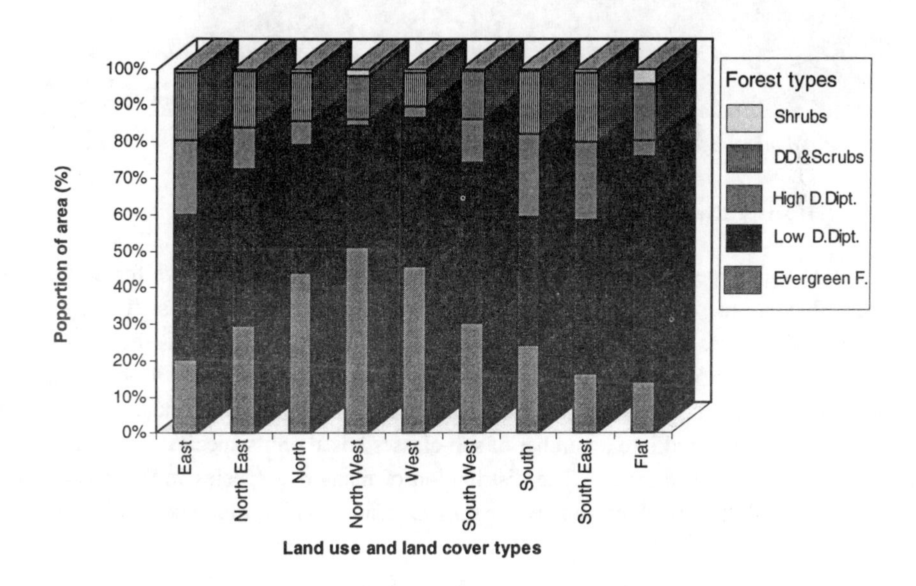

Fig. 78: **Relationship between land use/land cover types and the aspect of the study area III (Source: Author).**

In the case of the study area IV in the coastal zone, the study shows significantly good results. Except rangeland category all categories of the level I of the land use and land cover classification system shown in the Fig. 43 were derived in the Landsat classification of this study area. Both inland- and wetland vegetation ecosystem as well as the intensively managed land use systems dominate this study case.

Fig. 79: 3D perspective view of the portion of the study area IV (Source: Author).

There are two main types of forests, i.e. moist evergreen forest and mangrove forest, found in this area. They are the categories in level III of the land use/land cover classification system just mentioned above. Normally, both types are the end products of the land cover categorization in classifying a satellite image in many cases. In this study, it was found that mangrove forest could be classified into sub-classes using Landsat TM. Thus, the mangrove forest class was categorized into a number of sub-classes based on its species composition, soil conditions, and disturbance status. The distribution of mangrove species in this coastal zone concretely varies from shorelines and river banks to inland. This difference can be seen in a satellite image.

Distinguishing a rubber plantation was also possible in this study area. Both mature and young rubber plantation areas can be discriminated. Orchards are one of the main agricultural types in this region. There are various kinds of fruit trees in the orchard areas. A categorization orchard areas according to the species of fruit trees was not possible in this study. However, this class can be categorized into more sub-classes based on the difference of spectral values occurring in the class. It was found that the way of planting affected the difference of reflectance among the orchard areas. Thus, the classification was achieved by categorizing orchard areas into uniform planting and non-uniform planting. In addition, small-scale field crops were found among the orchards. The distinction of this land use type was also possible with some difficulties.

Paddy field were also distinguished in this case. Paddy fields occurred in this image were in the stage after harvesting. This class can be further categorized into two classes, i.e. harvested rice fields and harvested rice fields covered with grasses. Many paddy fields have been abandoned for a long time. These areas were covered with cordless grasses and aquatic plants. Distinction of swamps in wetlands was also successful in this classification. The reflectance of vegetation in this class is different enough from the reflectance of paddy fields. In settlement areas, the distinction of built-up areas was possible. Small villages which are not covered with tree crown closures can be also distinguished.

Shrimp farming areas were widely found in this coastal zone. This land use type can be easily delineated and subdivided into the level III of the classification system. In this area, shrimp farming areas vary from the site preparing stage (mangrove clearing) to completed shrimp ponds. From field surveys, it was found that some of the shrimp ponds were partly in use, while some of them have already been abandoned. The conditions of water and soil in these shrimp ponds presented a wide range of spectral response. Classifying this land use type was successful by supervised classification. Normally, this land use type can be delineated into different subclasses based on its components, such as water, mud, and levees. As stated before, the condition of water and soil in shrimp pond could vary greatly. Thus, further categorization was necessary in this study. Resulting subclasses have been grouped and defined to definite categories.

In conclusion, the level III of the land use and land cover classifications shown in the Fig. 43 were successfully derived using the original band of the Landsat TM data with the supervised classification algorithm. The transformed bands such as NDVI, Tasseled Cap, are not necessary to implement in this study area. Table 33 summarizes the land use and land cover categories distinguishable on Landsat TM data by computer-aided analysis method.

Tab. 33: Land use/land cover categories distinguishable on Landsat TM data by computer-aided analysis method for the study area IV

Level I	Level II	Level III	Level IV
Forest land	Evergreen forest	Moist evergreen	
	Mangrove forest	*Rhizophora spp.*	
		Rhisophora spp. intermixed with *Acevinia spp.*	
		Mixed mangrove along the bank of rivers	
		Mangrove found in elevated areas with harden mud	
		Mangrove clearing	

Tab. 33: (continued)

Level I	Level II	Level III	Level IV
Agricultural land	Orchards	Uniform fruit-trees Varying fruit-trees	
	Field crops	Small-scale crops	
	Rubber plantation	Mature rubber Young rubber	
	Paddy fields	Harvested rainfed rice fields	
		Paddy field covering with grasses and aquatic plants	
Barren land	Bare soil		
	Beach		
Urban area	Built-up areas		
	Open urban		
Wetland (non-forested)	Swamp (subject to sea water)		
Water	Shrimp ponds	Moderately turbid water	
		Muddy water to moisture soils	
	Inland ponds	(Clear fresh water)	
	Rivers and sea	Deep	
		Medium deep	
		Medium shallow and clear	
		Shallow and slightly turbid	
		Turbid or mud flat	

The proportion of land use and land cover types occurring in this area are summarized in Fig. 80. The figure illustrates that this study area is covered mostly with water and mangrove forest areas.

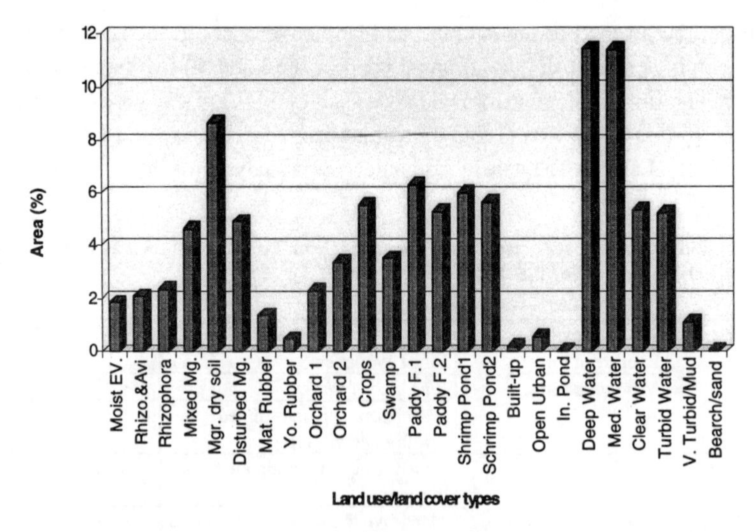

Fig. 80: Proportion of land use and land cover types in the study area IV (Source: Author).

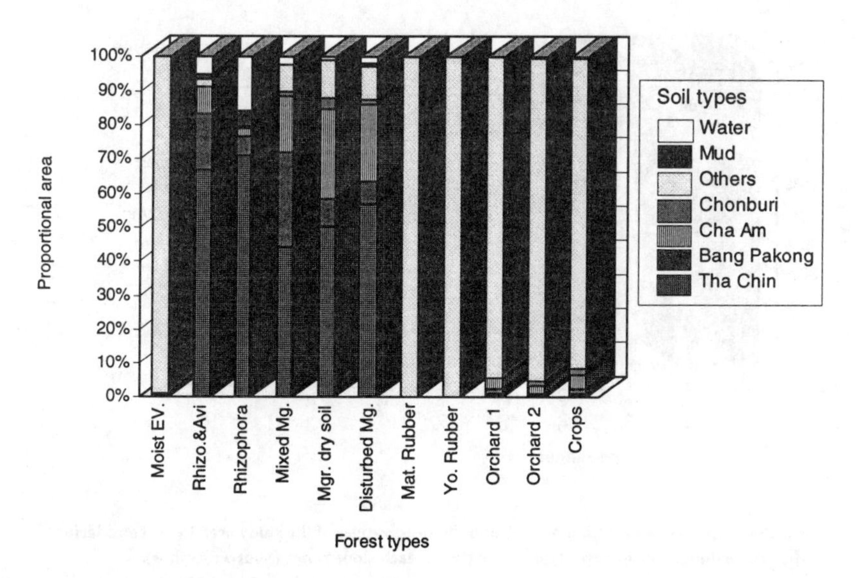

Fig. 81: **Relationship between the vegetation covers and soil types in the study area IV (Source: Author).**

SILAPATHONG (1992) pointed out that not all soil types found in this coastal zone were suitable for the distribution of mangrove forests. The study also stated that Tha Chin series (Tc), Bang Pakong series (Bpg), Cha Am series (Ca), and Chonburi series (Cb) are suitable for mangrove distribution. Thus, a matrix analysis was also carried out for this study area. The main forest types and also other vegetation types (categories) were analyzed in association with soil types. The result of this analysis is shown in Fig. 81. It can be noticed that the classification results are corresponding with the previous study mentioned above. All mangrove forest types have mostly been found on the Tha Chin series. Bang Pakong and Cha Am series are also suitable for the distribution. Furthermore, some small areas of mangrove forests, in particular mangrove in elevated areas and disturbed mangrove forest, were also found in other soil series. It can also be noticed that mangrove *Rhizophora* were also found in the water and mud. These observations can be essential for further studies in this coastal zone. Results and methodological techniques derived in this part of the study have been applied in this study area. This study case was documented in Chapter 7.

The analysis of the relationships between land use/land cover types and terrain data was also carried out. Main forest cover categories and agricultural cover categories were analyzed in association with slope ranges of the study area. Slope data were calculated using a DEM created in this study. Figures 82 and 83 illustrate this relationship. It can be seen that most of the land use/land cover type can be found in the slope gradient of 0-10%. Crops were found in the areas with slope range from 0 to 20%. Orchards were mostly found in the slope gradient of 10-20%. In the area with a slope of over 30%, only moist evergreen forest can be found. The results of this analysis corresponded with the observation from field surveys.

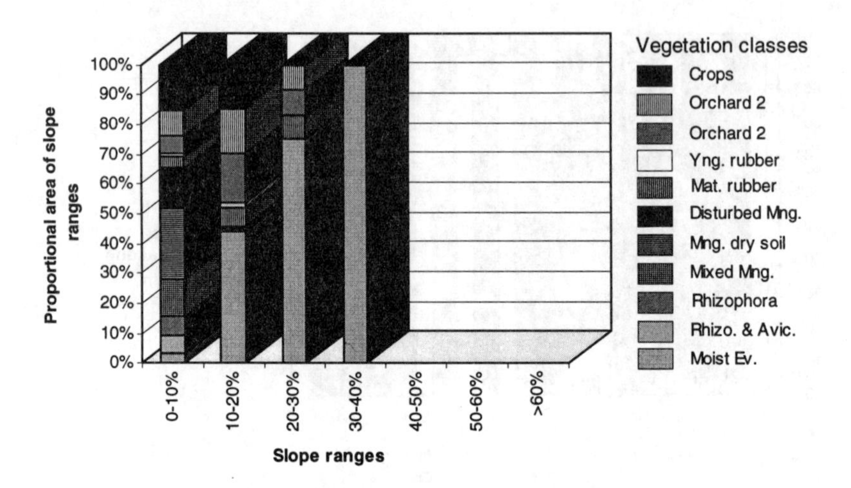

Fig. 82: **Relationship between vegetation cover and the slope ranges of the study area IV by considering the proportional area of cover types occurring in each slope range (Source: Author).**

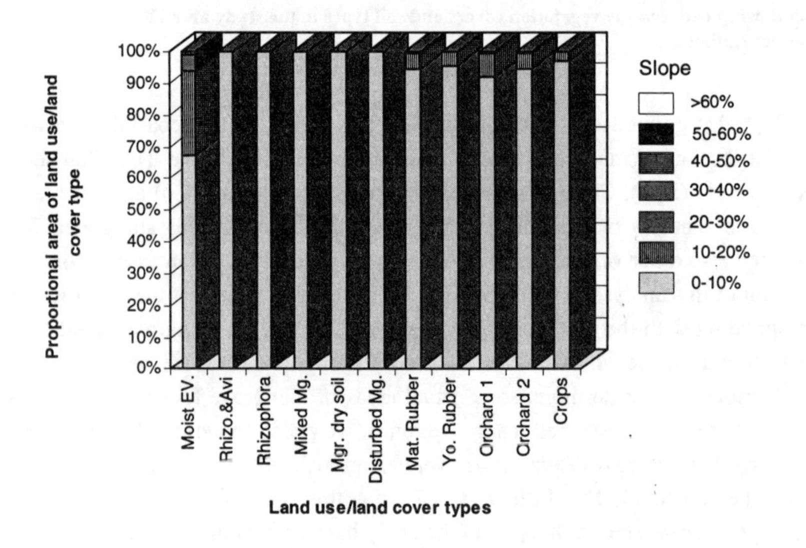

Fig. 83: **Relationship between vegetation cover and the slope ranges of the study area IV by considering the proportional area slope ranges occurring in each land cover type (Source: Author).**

6.2.2 CLASSIFICATION ACCURACY

An accuracy assessment was performed to define the classification results of each study area. Overall accuracy as well as the producer's and user's accuracy of each individual category were calculated for each set of analyzed data. The results of classification accuracy assessment are reported in forms of contingency tables as follows:

Tab. 34: Contingency table for the classification results of the study area I in the mountainous area (Source: Author).

Classi-fied data	Reference data											Total
	Wt	MD1	MD2	MD3	OC	DD1	DD2	DEv	PD1	PD2	Cr	
Wt	**-**											0
MD1		**23**	7					1			1	32
MD2		5	**20**	1				1		1	2	30
MD3			1	**61**		7	1				1	71
OC					**7**							7
DD1			1	1		**9**						11
DD2				2			**9**					11
Dev		1						**3**				4
PD1									**12**			12
PD2										**6**		6
Cr											**3**	3
Total	0	29	29	65	7	16	10	5	12	7	7	**187**

Abbreviations:

Wt = Water

MD1 = Dense mixed deciduous forest

MD2 = Disturbed mixed deciduous forest

MD3 = Mixed deciduous forest with bamboo

DD1 = Undisturbed dry dipterocarp forest

DD2 = Disturbed dry dipterocarp forest

DEv = Dry evergreen forest

PD1 = Crops in paddy fields

PD2 = Harvested paddy fields

OC = Old clearing

Cr = Upland crops

Reference data used in this error matrix are based on the data derived from the second ground survey.

Tab. 35: Classification accuracy of the study area I in the mountainous area (Source: Author).

Cover class	Producers accuracy (%)	Users accuracy (%)
Wt	-	-
MD1	79.3	71.9
MD2	69.0	66.7
MD3	93.8	85.9
OC	100.0	100.0
DD1	56.2	81.8
DD2	90.0	81.8
Dev	60.0	75.0
PD1	100.0	100.0
PD2	85.7	100.0
Cr	42.9	100.0

Overall Classification Accuracy = 81.82 %

Tab. 36: Contingency table for the classification results of the study area II in the Central Plain (Source: Author).

Classi-	Reference data								Total
fied data	Wt	Min	PF	hPF	Sc	Tr	Bu	opU	
Wt	**7**		2						9
tWt		**1**							1
PF	3		**126**		3			3	137
hPF				**30**	6				6
Sc			5		**4**			9	48
Tr			1			**30**			31
Bu							**1**		1
opU				1				**13**	14
Total	10	1	134	33	13	30	1	25	**247**

Abbreviations:

Wt = Water
Min = Sand mining areas
PF = Paddy fields
hPF = Harvested paddy fields
Sc = Sugarcane
Tr = Trees, home gardens
Bu = Built-up areas
opU = Open urban

The reference data used in this error matrix is based on the data derived from the second ground survey. Only the main land use classes were used to assess the accuracy due to the difficulty of deriving validate reference data.

Tab. 37: Classification accuracy of the study area II in the Central Plain (Source: Author).

Cover class	Producers accuracy (%)	Users accuracy (%)
Wt	70.0	77.8
Min	100.0	100.0
PF	94.0	92.0
hPF	90.9	100.0
Sc	46.2	62.5
Tr	100.0	96.8
Bu	100.0	100.0
opU	52.0	92.9

Overall Classification Accuracy = 86.64 %

Tab. 38: Contingency table for the classification results of the study area III in the Korat plateau (Source: Author).

Classifi-ed data	Reference data													To-tal
	Ev	LD	HD	Scr	Sc	DW	SW	TW	PF	PF/U	Vi	BS	Cr	
Ev	**12**	2				2								16
LD		**24**		1										25
HD		3	**5**											8
Scr	1			**9**										10
Sc					**14**									14
DW						**3**								3
SW							**7**							7
TW								**1**						1
PF		1		2				1	**89**	4			1	95
PF/U		1							2	**49**				54
BS									1		**10**			12
Vi												**1**		1
Cr									1				**2**	3
Totals	13	31	5	12	14	5	7	2	93	53	10	1	3	**249**

Abbreviations:

Ev = Evergreen forest	DW = Deep water
LD = Low dry dipterocarp forest	SW = Shallow water
HD = High dry dipterocarp forest	TW = Turbid water
Scr = Scrubs (Dry dipterocarp)	PF = Harvested paddy fields
Sc = Sugarcane	PF/U = Harvested paddy fields in undulating areas
Cr = Other crops	
Vi = Villages	BS = Bare soil

The reference data used in this error matrix is also based on the data derived from the second ground survey.

Tab. 39: Classification accuracy of the study area III in the Korat plateau (Source: Author).

Cover class	Producers accuracy (%)	Users accuracy (%)
Ev	92.3	75.0
LD	77.4	96.0
HD	100.0	62.5
Scr	75.0	90.0
Sc	100.0	100.0
DW	60.0	100.0
SW	100.0	100.0
TW	50.0	100.0
PF	95.7	93.7
PF/U	92.5	90.7
Vi	100.0	83.3
BS	100.0	100.0
Cr	66.7	66.7

Overall Classification Accuracy = 90.76 %

Tab. 40: Contingency table for the classification results of the study area IV in the coastal zone (Source: Author).

*	Reference Data																			To-
	Ev	RA	Rz	mM	Mh	Mc	mR	yR	O1	O2	Cr	Sw	PF	Shr	Bu	oU	inP	Wt	Md	tal
Ev	**4**								1											5
RA		**4**	1																	5
Rz			**6**																	6
mM				**11**																11
Mh					**17**									3		1				21
Mc						**7**								5						12
mR							**3**													3
yR								**1**												1
O1				1					**3**	2										6
O2							1			**7**										8
Cr				1						3	**10**									14
Sw												**8**		1						9
PF					1						1	1	**25**							28
Shr														**29**						29
Bu															**1**					1
oU													1			**-**				1
inP																	**1**			1
Wt																		**80**	1	81
Md																			**2**	2
Total	4	4	7	13	18	7	4	1	4	12	11	9	26	38	1	1	1	80	3	**244**

Abbreviations:

* Classified data	Cr = Crops
Ev = Evergreen forest	Sw = Swamps
RA = Mangrove Rhizophora & Avicenia	PF = Paddy fields
Rz = Mangrove Rhizophora	Shr = Shrimp ponds
mM = Mixed mangrove	Bu = Built-up areas
Mh = Mangrove on dry soil	oU = Open urban
Mc = Mangrove clearing	inP = Inland pond
mR = Mature rubber plantation	Wt = Water
yR = Young rubber plantation	Md = Mud
O1 = Orchard with uniform fruit-trees	O2 = Orchard with varying fruit trees

Reference data used in this error matrix is also based on the data derived from the second ground survey, topographic maps, and available thematic maps from previous study.

Tab. 41: Classification accuracy of the study area IV in the coastal zone (Source: Author).

Cover class	Producers accuracy (%)	Users accuracy (%)
Ev	100.0	80.0
RA	100.0	80.0
Rz	85.7	100.0
mM	84.6	100.0
Mh	94.4	81.0
Mc	100.0	58.3
mR	75.0	100.0
yR	100.0	100.0
O1	75.0	50.0
O2	58.3	87.5
Cr	90.9	71.4
Sw	88.9	88.9
PF	96.2	89.3
Shr	76.3	100.0
Bu	100.0	100.0
oU	0.0	0.0
inP	100.0	100.0
Wt	100.0	98.8
Md	66.7	100.0

Overall Classification Accuracy = 89.75 %

These tables show promising results. It can be concluded that the utility of transformed TM data also appears to be very highly successful for natural resources classifications in Thailand. It can be more widely used in various applications.

6.2.3 PROSPECTS FOR FURTHER STUDIES

I Classification strategies for classifying some land covers on Satellite image

In addition to the image classifications mentioned before, there are some additional techniques applied to discriminate land cover classes that should be noted. These are:

(1) Strategies for classifying settlement areas

In different regions of Thailand, the settlement areas are different in pattern. These features may characterize different land covers in a satellite imagery. Built-up areas and urban lands constitute a spectrally heterogeneous land use class, while the pattern of rural settlement areas in different regions are commonly sparse. As an example in rural areas of the study area IV in the coastal zone, the settlement is commonly a piece of land with a house and some cultivation including flower plants and unproductive trees. Crops found around each single house are such as coconuts, mangoes, bananas, vegetables, etc. In the study II in the Central Plain, housing is isolated and commonly surrounded with a big home garden or an orchard. The building is usually hidden under a clown closure of big trees. As a result, it is generally difficult to discriminate such settlement areas from satellite. Only the built-up areas can be discriminated. The discrimination of settlement areas in the study area of the Korat plateau is possible by applying a simple technique. The discrimination was successful in examining the reflectance properties of every pixel in that area. In other words, training for land use classes of villages in the supervised classification was not based on the isolation reflectance of housing pixels, but based on neighboring pixels that can be defined as the vegetation cover of home gardens surrounding the houses. Fortunately, the accuracy for the distinction of this class was excellent, although a few pixels of other class is appearing like seed, were produced during supervised statistical pattern recognition in the classification process, these isolated pixels could be eliminated using filtering algorithm in the post-classification process.

(2) Strategies for delineating roads on satellite images

Since roads are one of the important features in mapping, the extraction of roads as well as other similar linear networks have to be done in the study. Topographic maps used as the base maps in the study are relatively out of date, road lines on these maps are not suitable to be transfered to land use/ land cover maps being produced in this study. Due to the high spatial resolution of Landsat TM satellite imagery, the extraction of such features using satellite imagery has become more feasible than before. In any case, these linear features can be made visible from a classification because of their continuity and their spectral homogeneity within their surfaces which are generally quite contrast to the other surrounding land covers. Hence, the extraction of these features by manual delineation from an image were done to implement the road alignments when they were missing.

In fact, the road delineation from satellite imagery depends on the image quality, the surface of road, and the complexity of road networks. Roads in urban areas, rural areas, or in mountainous areas have different characteristics. Furthermore, roads can have different outlines varying from straight lines to partially occluded line structures. In this study, various kinds of roads as well as other linear networks occur on all satellite images used. For example, in the study area I in the mountainous area there are both major roads and forest roads. In the study area II in the Central Plain there are very complex of different type of roads and irrigation canals. In the study area III, there are also different types of roads varying from trails to highways. In the study area IV in the coastal zone, the complexity of roads also occurred. Three steps were employed for the task of delineating linear networks in this study. First, associated knowledge was applied considering the result of the image classification based on the land-cover. Second, the enhancement of original bands of satellite data, such as filtering with edge detection, images transformations, were carried out, and then the results were use to extract linear features. Third, with help of the Digital Elevation Model, 3D perspective views were displayed, specific bands of a satellite image and an enhanced image were then draped on the 3D terrain models. Viewing of the 3D terrain models could also help in identify linear structures occurring on images. The third and the second steps were generally done together, while the digitizing approach was employed during this operation. However, these operations can be done simultaneously only on a Unix workstation system.

II Comments for using satellite data for land use/land cover classification in Thailand

It can be concluded that the Landsat data analysis for land use and land cover classification in this study yielded overall satisfying results that can be used as reference for further studies. Landsat TM data have been shown to provide an effective way of deriving environmental information both at a regional and a specific site level in various environments. It is a data source for many study purposes. In Thailand, the use of such a data is usually related to either forest land management or agricultural land management. For forest classification and monitoring, it should be noted that homogeneous canopies can be classified and distinguished without difficulty. Heterogeneous canopies such as mixed deciduous forest may need to be classified into more classes. These classes can be grouped to derive a satisfying category. It is difficult to classify stands on steep western slope because of the shadows cast during the orbit of Landsat in the morning. However, techniques for integrating a DEM or masking can be used to reduce the shadow effect. Use of Landsat TM data is also practicable to survey the mangrove forests in coastal zones. Due to the fact that coastlines are the critical spatial characteristics in every coastal zone, the delineation coastlines from Landsat data is possible. However, the coastlines from topographic map should also be considered. The Digital Elevation Model can be integrated to the coastal zone study both for modeling a terrain or for modeling a depth in the sea.

In the case of studies related with agricultural land use, Landsat TM data can be adopted in many cases. An evaluation of Landsat data for agriculture and land use can be made using an hierarchical approach. In the approach, agricultural lands can be subdivided into categories needed by considering the possibility of discrimination those categories from image. If there is vegetation found in an agricultural land, the categorization can be based on the consideration of:

- its type i.e. temporary crop or permanent crop
- its physiognomy, by considering growth status, crown closure density, or the planting patterns.
- its habitat (e.g. riverine vegetation, upland cropping)

Cropping areas can be coarsely categorized as temporary crops and permanent crops. Temporary crops normally do not appear all year round. Thus, an actual crop calendar is essential. Paddy fields are one of the main land use types that can be found in the whole country. The status of paddy fields can vary from one region to the other. Thus, a crop calendar is also essential in the satellite imagery classification. The reflectance values in a paddy field can vary greatly owing to the composition of objects in the field. Development stages of rice plants and soil conditions in relation to the Landsat date commonly yields the overall spectral values of the paddy fields. As a result, the classification of Landsat TM data to discriminate the paddy fields can yield unsatisfactory results. Thus, the classification of this land use type needs the understanding how classifiers assign a class in a computer-aided classification system. Classifying the paddy fields into sub-classes may have to be done, and then these sub-classes would be grouped into one class desired.

In addition to the Landsat image classification, factors affecting the image data have to be considered. These factors are commonly related with the environmental change of the interesting area. Examples of the changes in forested areas are such as: changes from deforestation, changes from erosion, changes from fire, seasonal changes in the case of deciduous forests. Examples of the changes in agricultural areas are such as: time of growing, site preparation or harvesting, changes from irrigation or drought, changes from human-treatments.

III Tools for planning

This study has provided information charts which can be used as a decision rule for further studies relevant to satellite data analysis and field work (Fig. 84 to 87). These charts were created based on the consideration of various information types derived from this study. The information sources are derived from the climatological data, crop calendars and general features of Thailand.

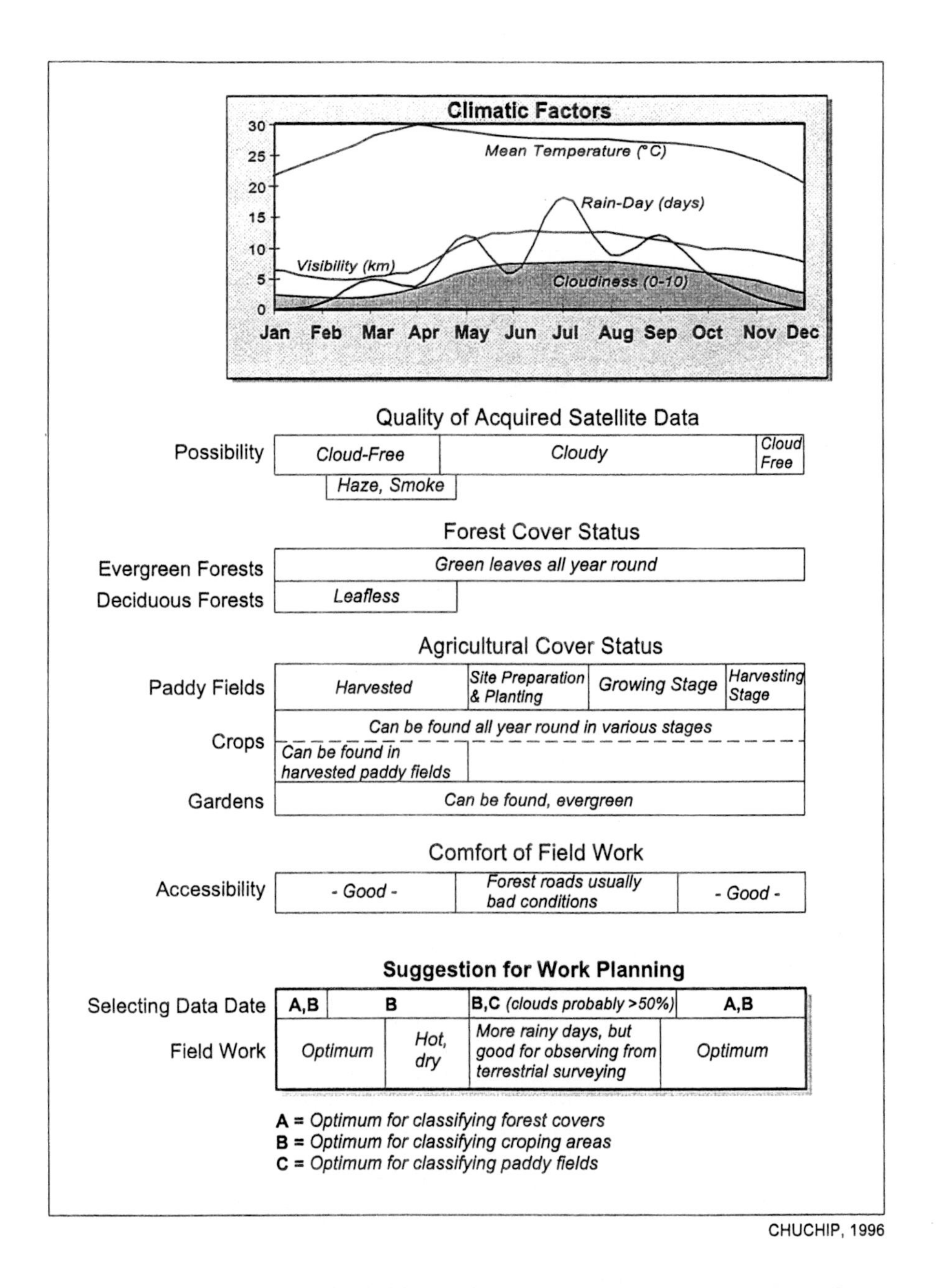

CHUCHIP, 1996

Fig. 84: **Information chart for decision making in using satellite data in the study area I in the mountainous area and neighbouring regions (Source: Author).**

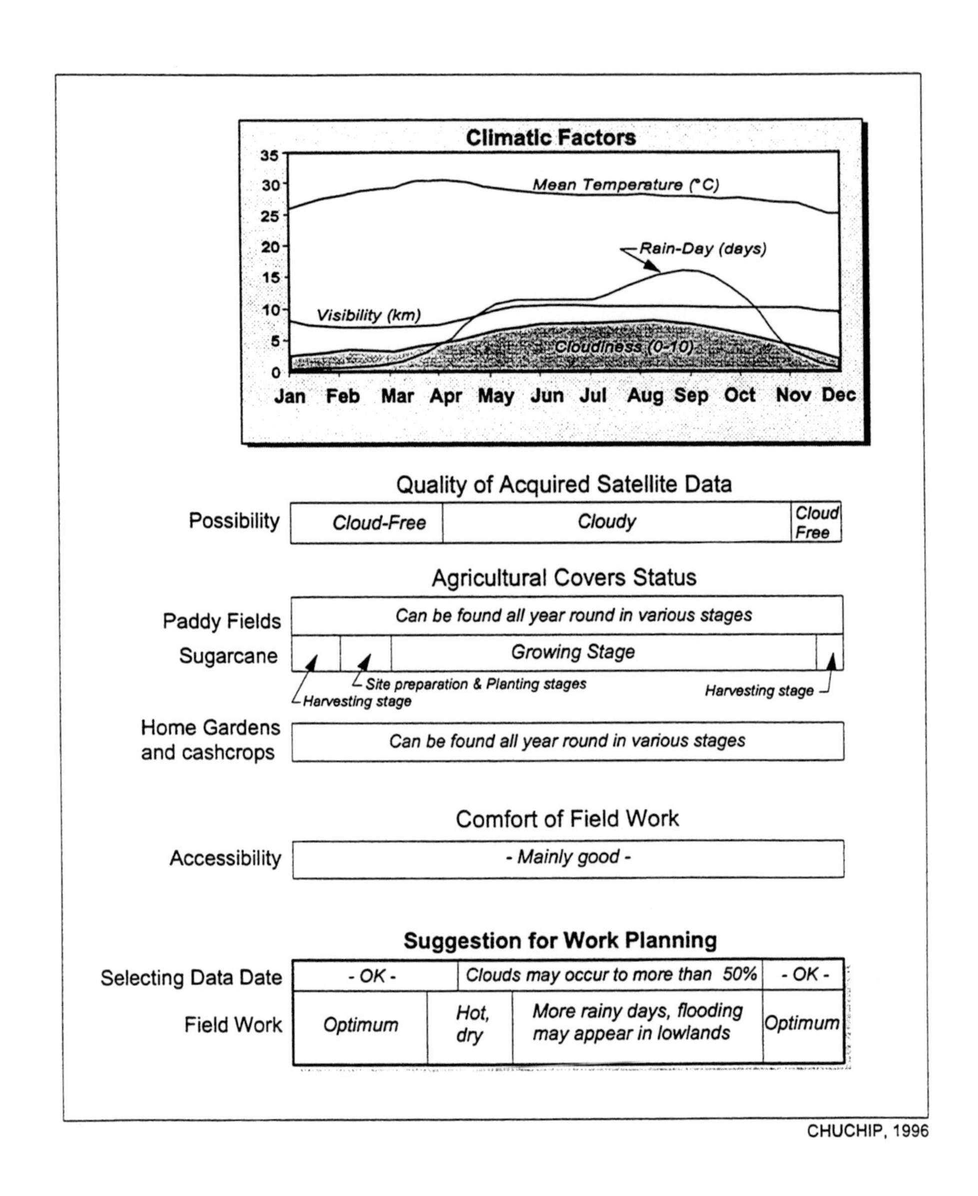

Fig. 85: Information chart for decision making in using Satellite data in the study area II in the Central Plain and neighbouring regions (Source: Author).

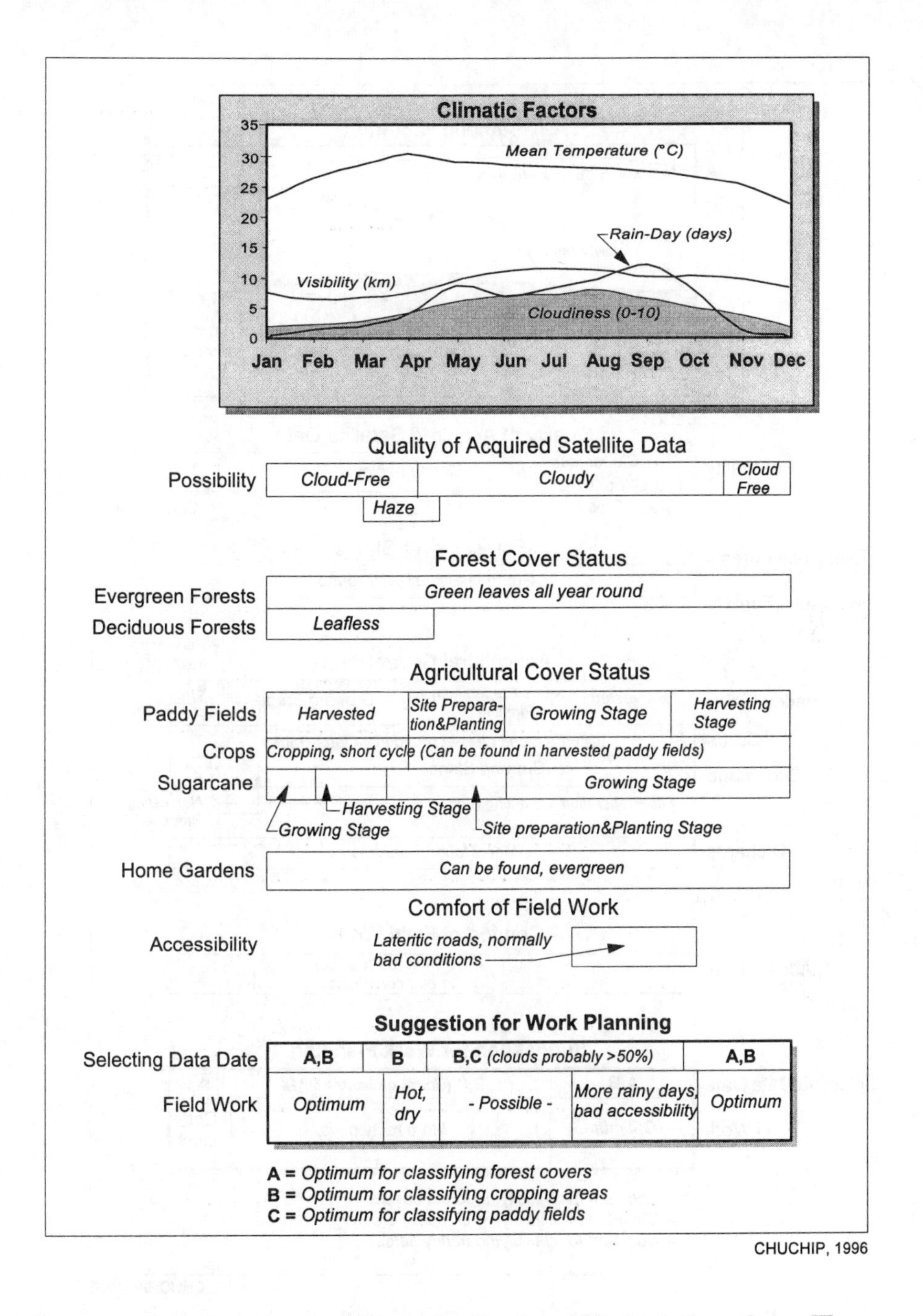

Fig. 86: **Information chart for decision making in using satellite data in the syudy area III in the Korat plateau and neighbouring regions (Source: Author).**

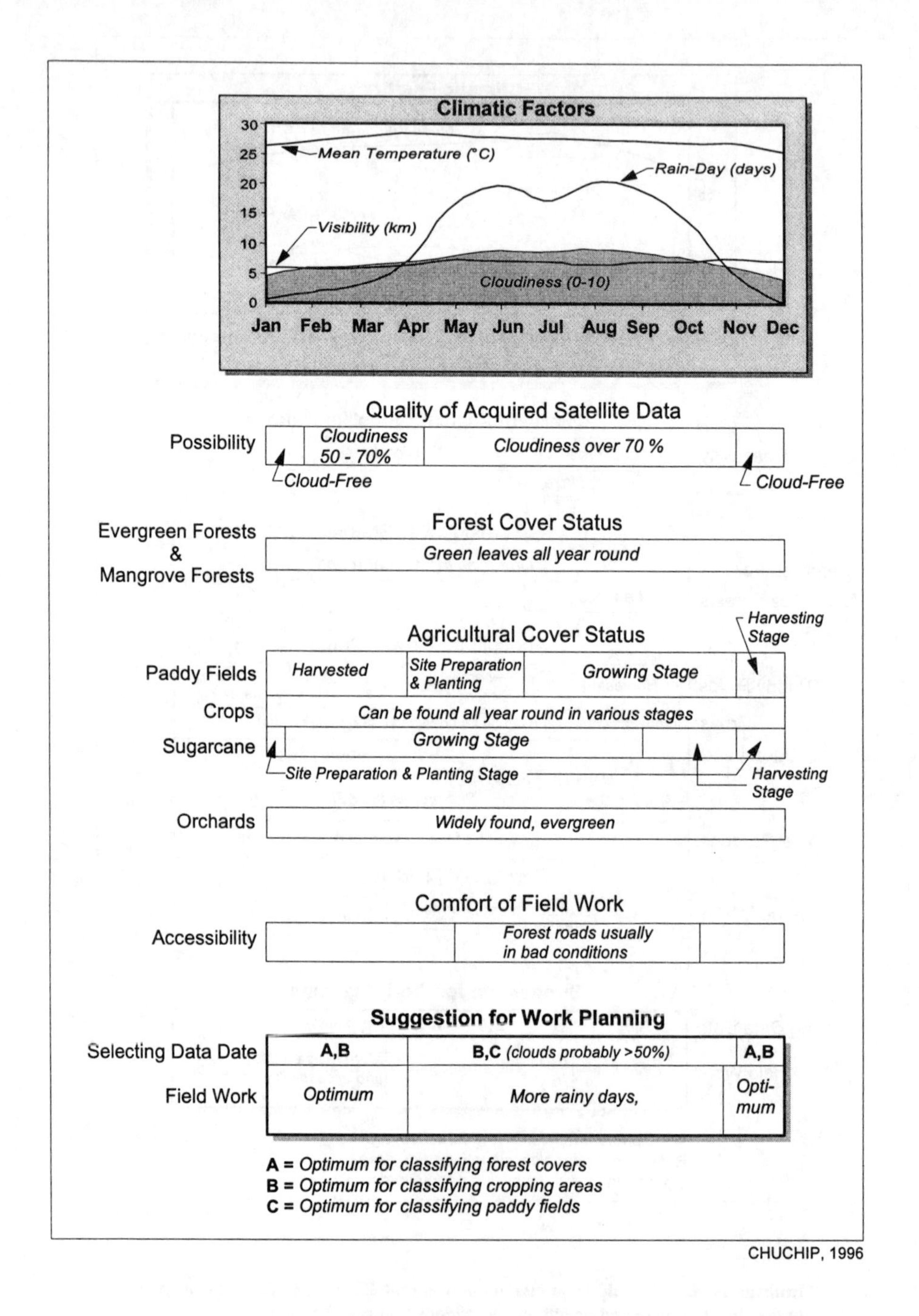

Fig. 87: **Information chart for decision making in using satellite data in the study area IV in the coastal zone and neighbouring regions (Source: Author).**

6.3 Optimized land use and land cover classification system

The classification of the four study areas show overall satisfying results. However, some of land cover units identified from the first field survey (Tab. 22, 23, 24 and 25) could not be categorized into discrete spectral classes, such as trails, dikes and ditches of a shrimp pond for instance. Thus, the hierarchical land use and land cover classification system developed earlier (Fig. 43) was reorganized again. This hierarchical system is shown in Fig. 88.

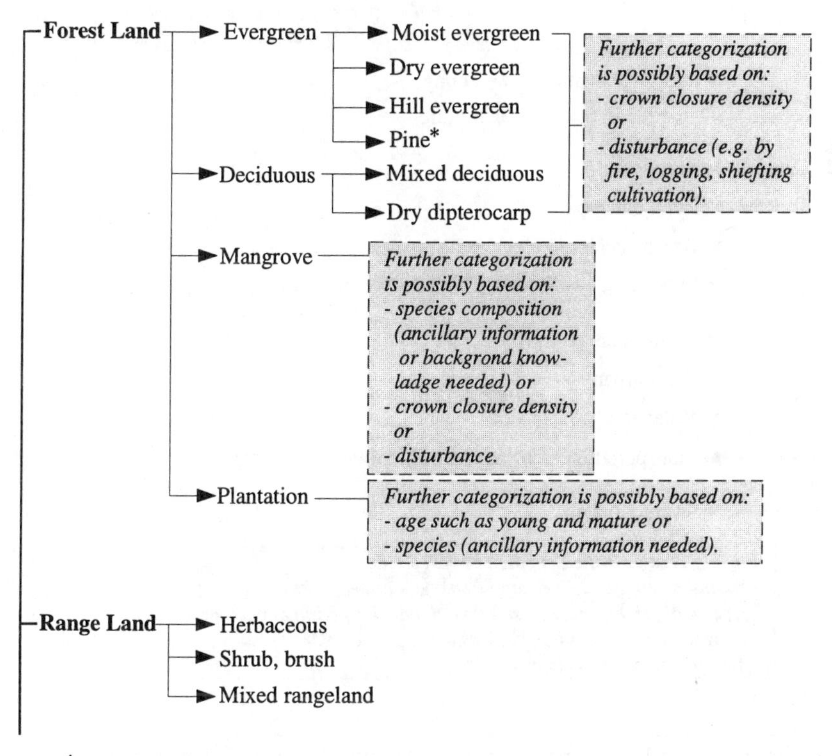

- continue next page -

Fig. 88: The hierarchical land use and land cover classification system for use with Landsat TM data by means of computer-aided analysis system (Source: Author).

Fig. 88: **(continued)**

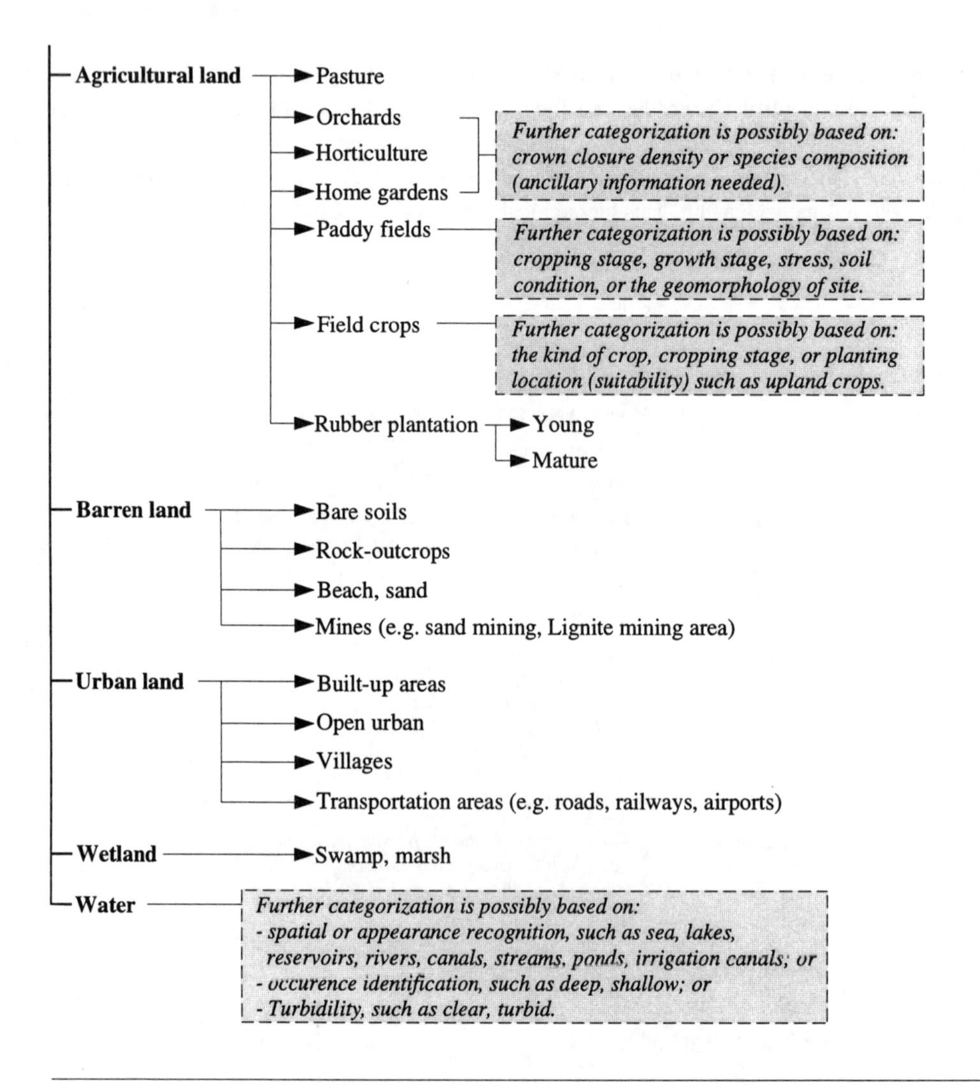

Remark: * Pine forest usually occurring on mountains (normally above 1000 m. MSL) in the nothern region of Thailand should be included in this level.

It should be noted that a complete subdivision of all categories may not be essential in some cases. This study was aimed at showing the possibility of the differentiation of land use and land cover types in the sense of high resolution satellite data analysis. The system also provides the flexibility of modifications in order to support a wide range of further applications. For future applications, categories within the classification system can also be aggregated. In such cases, lower level categories can be grouped together in order to provide data for more general categories as needed. However, this system is compatible at more generalized levels with previously developed classification systems.

In order to support the inventory of land resources in the future and to support the data management of a computer-based Geographic Information System, the hierarchical system shown in the Fig. 88 was reorganized and coded. Numerical codes were given to each category. The number of digits reflects the level of detail in the classification system being used. The first digit identifies the level I class, the second digit identifies the level II class, and the third digit identifies the level III class. These numbers can be essential for coding land use/land cover data into a GIS database. Numerical codes can be used as identification number (ID) in the attribute data of the GIS system. This land use and land cover classification system is suitable for use with Landsat TM data by means of a computer-aided system.

6.4 Land use and land cover maps of the study areas

Results of the image classification of each study area by means of computer-aided mapping systems are shown in Appendix E (4 thematic maps attached in the back cover of this book). It has to be remarks that the land use and land cover map of the study area IV in the coastal zone is based on the classification done in the second part of the study (Chapter 7).

Tab. 42: **Land use and land cover classification system for use with Landsat TM data by means of computer-aided analysis (Source: Author).**

Level I	*Level II*	*Level III*	*Remarks*
100 Urban Land	110 Built-up area 120 Open urban 130 Villages 140 Transportation areas	141 Roads 141.1 Asphalt 141.2 Laterite 142 Railways 143 Airports	
200 Agricultural Land	210 Orchards 220 Home Gardens	xxA *(A=1 or 2 when 1= Uniform 2= Varying)* xxA.B *(B= Any number referred to the species ID)*	*Further categorization at level III is based on:* *A. Crown closure density or/and* *B. Species composition.* *Species ID being used for further categorization:* *1. Citrus 7. Custrad* *2. Durian 8. Jujube* *3. Rambutan 9. Coconut* *4. Longan 10. Tamarind* *5. Litchi 11. Banana* *6. Mango 12. Other*
	230 Horticulture		
	240 Paddy Fields	xxA *(A=1 to 4 when 1= Site preparation 2= Young stage 3= Mature stage 4= Harveting stage* xxB *(B= 5 to 7 when 5= High water 6= Wet soil 7= Dry soil* xxC *(B= 8 to 9 when 8= Low flat plain 9= High flat plain*	*Further categorization at level III is based on:* *A. Cropping and growth stages, or* *B. Soil and water conditions, or* *C. The geomorphology of site (in case of harvested paddy fields).*
	250 Field Crops	xxA *(A can be specified the same number as paddy fields)* xxA.B*(B= Any number referred to ID of the kind of crop)* xxC *(C= 5 or 6 when 5= Lowland 6= Upland)*	*Further categorization in the level 3 is based on:* *A. Cropping stage, or/and* *B. The kind of crops, or* *C. Planting site suitability.* *Crop ID being used for further categorization:* *1. Sugarcane 6. Kenaf* *2. Cassava 7. Tobacco* *3. Soyabean 8. Corn* *4. Peanut 9. Vegetation* *5. Pineaple 10. Other*
	260 Rubber Plantation 270 Pasture	261 Young 262 Mature	

Footnote: '0' is used to indentify the level class that is not specified.

- continue next page -

Tab. 42: (continued)

Level I	*Level II*	*Level III*	*Remarks*
300 Range Land	310 Herbaceous 320 Shrub, brush 330 Mixed		
400 Forest Land	410 Moist Evergreen 420 Dry Evergreen 430 Hill Evergreen 440 Pine 450 Mixed Deciduous 460 Dry Dipterocarp	xxA *(A= 1 or 2 when 1= Undisturbed 2= Disturbed* xxB *(B=3 or 4 when 3= Densed 4= Opened)*	*Further categorization at level III is based on: A. Disturbance, or B. Crown closure density.*
	470 Mangrove	xxA *(A= 1 or 2 when 1= Undisturbed 2= Disturbed* xxB *(B=3 or 4 when 3= Densed 4= Opened)* xxC *(C=5, 6, or 7 when 5= Rhizophora spp. 6= Rhizophora spp. mixed with Acivenia spp. 7= Mixed*	*Further categorization at at level III is based on: A. Disturbance, or B. Crown closure density, or C. Species composition.*
	490 Plantation	xxA *(A= 1 or 2 when 1= Young 2= Mature* xxB *(B=3, 4, or 5 when 3= Teak 4= Pine 5= Others*	*Further categorization at level III is based on: A. Age B. Species. (Either xxA.B or xxB.A can be further applied.)*
500 Barren Land	510 Bare Soil 520 Rock-outcrop 530 Beach 540 Mines	541 Lignite mining 542 Others	
600 Wetland	610 Fresh water swamp 620 Salt water swamp		
700 Water	710 Streams (Huai) 720 Canals 730 Rivers 740 Ponds 750 Lakes 760 Reservoirs 770 Gulfs	xxA *(A= 1 to 4 when 1= Clear 2= Slightly turbid 3= Turbid 4= Muddy*	*Further categorization at level II is based on: - spatial or appearance and can be further categorized for the 3rd level based on: - turbidility or occurance.*

Footnote: '0' is used to indentify the level class that is not specified.

CHUCHIP, 1996

PART II APPLICATION

7 MANAGEMENT AND APPLICATION OF THE STUDY RESULTS IN A CASE STUDY

7.1 GIS database establishment

The first step in any forms of automated geographic data processing consists of the transformation of analog models of objects on the Earth's surface into machine readable formats. Conceptually, the process of map digitization is an exercise in transferring these objects into machine readable format. Another source of geographic data are the direct captures of images of the Earth. In other words, geographical entities can be captured from maps or images and subsequently represented as points, lines, polygons, or a matrix of numbers. Most questions in data capture are related to scale, resolution, and the efficient storage and retrieval of the spatial entities with respect to the ultimate use of the data. Figure 89 shows the chart of the structure of geographic data sets of the study area.

In this study, geographic data were organized and modeled by means of the ARC/INFO concepts. There are three main sources of data that were adopted to produce important spatial information used in this study. The first are available thematic maps, i.e. topographic maps, soil maps, hydrological maps, geological maps. The second are images derived from remotely sensed data, i.e. classified images of land use and land cover types. The third are topographic data derived from DEMs or TINs, such as aspect, slope, elevation. These data contained within those sources are described graphically as sets of points, lines, surfaces or areas. With ARC/INFO, these data were extracted by digitizing and then transformed as vector or raster data depending on the purposes of further use and convenience. To store vector data, *coverages* were formed. Feature attribute tables were then built for each *coverage*. Normally, vector coverages are suited for forming the cartographic and feature database within a GIS while cell-based data, *grids*, provide the ability of modeling, manipulating and analyzing geographic data. *Grids* were also adopted here to perform for each data set.

In addition, image data were also taken into account of this GIS database because they can be used as map displays as well as attributes that describe spatial features. These images are such as the band combination of satellite image, scanned topographic maps, diagrams of sample plots, pictures of terrains or sample plots from field surveys, etc.

7.2 GIS analysis in a case study of the study area IV : An example of data integration

An attempt of analyzing the data and modeling a GIS from the database was pointed out in order to demonstrate that the methodology used in this study can be applicable for further similar case studies in Thailand. The study area IV in the coastal zone was selected to demonstrate here, because its critical environmental features is interested.

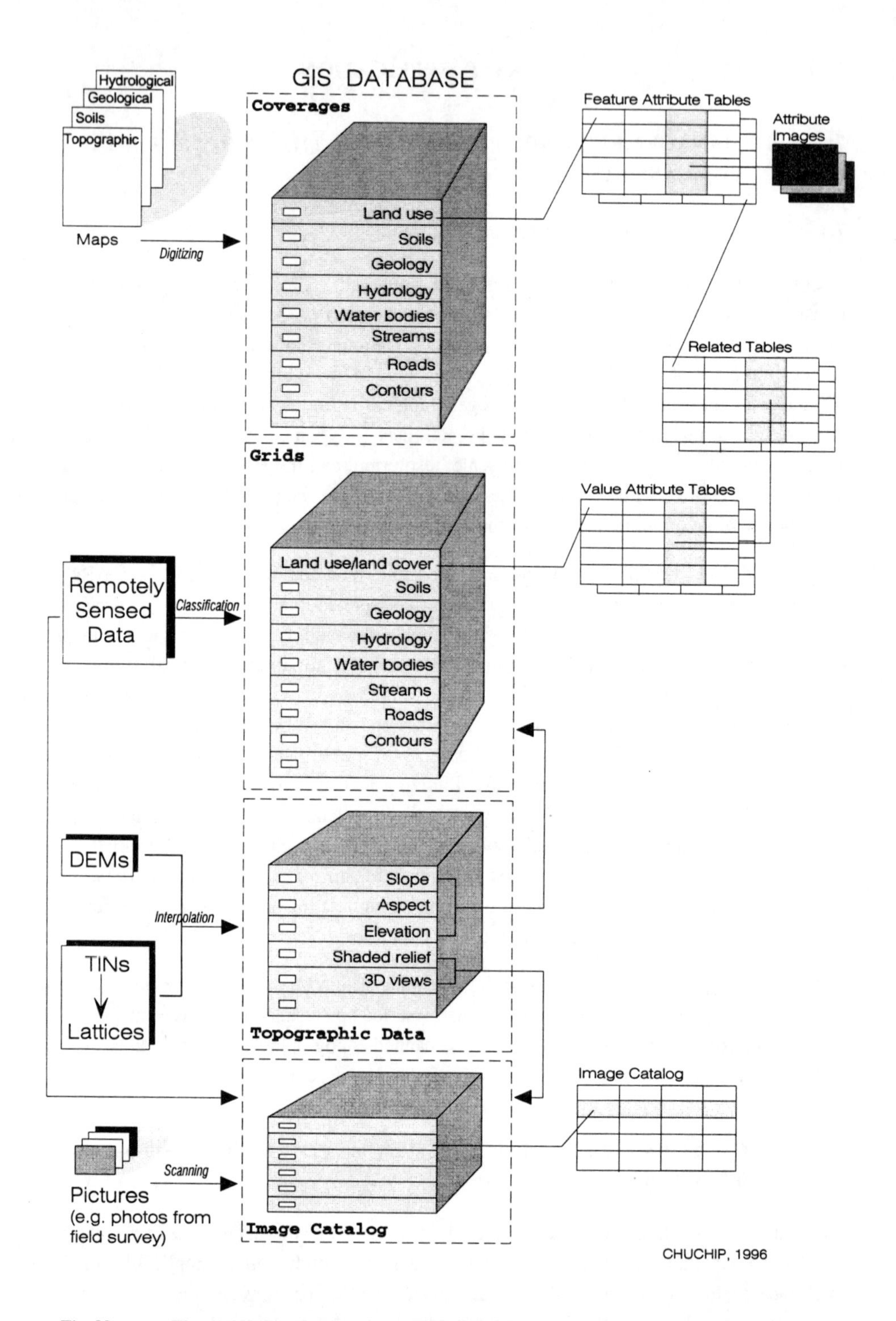

Fig. 89: **Flowchart of a GIS database established in this study (Source: Author).**

Coastal zones with their mangroves and the continental shelf are biologically the most productive marine environments in Thailand. The coastal environment is complex and fragile. The status of the coastal zone in Thailand is usually under intensive pressure from over-fishing, illegal shrimp farming and mangrove depletion because of population growth. Some zones are also affected by other activities such as waste discharge, mining, onshore activities related to agriculture, industry, or deforestation. Thus, there is an urgent need to improve assessment and monitoring of the condition of coastal zones. Unfortunately, available data are usually scarce, outdated, based on old maps and aerial photographs, or derived from ground enquiries likely to be incomplete and of doubtful reliability. Fortunately, the operational availability of high resolution satellite imagery, namely Landsat Thematic Mapper, and SPOT, opens new possibilities for investigating and monitoring coastal resources. Compared with information acquired by traditional methods, these data offer a number of advantages. Thus, the main purpose of this case study was to assess the technical possibilities of integrating satellite data and surface modeling with GIS for land use planning in the coastal zone. The test site is part of Chantaburi province in the East of Thailand. Here, the aquaculture is widely developed and the coast is covered with mangrove forests.

There are a few number of studies associated with this area. For instance the one is the study of SILAPATHONG (1992). This study also applied a GIS for mangrove forest management in this area. The study showed that most mangrove forest areas in this region have been exploited through both legal and illegal shrimp farming and logging. Around 50 percent of the total mangrove forests had been depleted in the past decade. The study pointed out that the mangrove forests were closely correlated with the physical factors of the site, i.e. altitude, soil type, and water salinity. An elevation of 7.5 m is the upper limit of mangrove forest distribution. A low tide of 0.6 m and a high tide of 2.5 m are the tidal zone of those mangrove forests. There are only some soil series that are suitable for the mangrove forest distribution, i.e. Tha Chin series, Cha-am series, Chonburi series, and Bang Pakong series (The soil descriptions are in the section 4.4 of this book). Sea water in this zone having 20-27 ppt. salinities shows a suitability for mangrove forest distribution.

With the up-to-date information derived from the second field trip in Thailand during September to November 1994, it was found that the Royal Forest Department has been monitoring, mapping and managing this mangrove forest zone. 100 m from the coastline and 20 m from the bank of canals or rivers have been legally announced to be reserved as the conservation zone. However, it seems to be difficult in practice. During this field trip, it was found that many mangrove forests in the conservation zone have were severely depleted. The conflict between the foresters and the people in the area is very evident. The local forest protection station has been trying to reforest in the area of the conservation zone and even in the tidal zone towards the shorelines in order to avoid conflicts with the farmers who occupy the areas in this zone. Reforestation in the water areas seems to be successful but not in all cases. Mangrove seedlings were partially damaged due to being flooded under water for a long time during the high tide, in particular where the seabed is too deep. Hence, an example of the practical use of data from the established GIS database and relevant techniques was then carried out in this study. The technique of surface modeling in the water area of this zone was tried out to locate depth curves that are critical for mangrove plantation under water.

With available information on the established database (from Section 7.1) and the results of SILAPATHONG (1992), the methodology of mapping a mangrove forest potential map in particular for the conservation zones was performed. Actually, SILAPATHONG has also provided a potential map for some parts of this test site. In this study, the potential mapping was done with different techniques and purposes. Furthermore, the intertidal zone of this region was considered in this study as a critical zone. Areas from coastlines to sea was included in the analysis.

To do this research, four aspects have been concerned; i.e. the status of the existing land uses and land covers, the physical criteria for the Khlung mangrove forest site potential, buffer zones from coastlines and streams. In addittion, more new satellite data, i.e. Landsat TM and SPOT XS, have been used in this stage. These satellite data were used to improve the result of the land use/land cover classification derived before. Table 43 shows the important information of these satellite images.

Tab. 43: Technical information of satellite data used in this stage (Source: Author).

Information	Satellite Products	
	Landsat TM	SPOT XS
Product Type:	BULK	BULK
Product Location	Scene Center: Lat.: 13.005 Long.: 101.962 (Path/Row: 128/51)	Scene Center: Lat.: 12.519 Long.: 102.141 (Path/Row: 266/325)
Product Scene Center Time	01/02/1993; 02:53:48	12/02/1993
Sun Elevation at Product Center: (degrees)	41.94613	55.11270
Sun Azimuth at Product Center: (degrees)	127.75336	137.76994
Sensor Pointing Angle: (degrees)	0.00	2.60
Input Volume:		
Medium (tape type):	High Density	High Density
Orbit Number:	47451	223
Signal Acquired at:	02:51:00.0000	03:49:27.0000
Signal Lost at:	02:56:00.0000	03:53:31.0000
Total Number of Swaths:	359	300
Swaths with Sync Losses:	0	0
Total number of sync losses:	0	0
Line Length Variation:	6314.1 - 6324.1	-
Radiometric Options:		
Radiometric:	CAL2	CAL2
Representation:	Linear	Linear
Geometric Options:		
Correction type:	Bulk	Bulk
Resampling kernel:	DAMPED 8 POINT SINC	DAMPED 8 POINT SINC
Elevation Correction:	Not applied	Not applied
Resolution (pixel size)	30 x 30 m	20 x 20 m
Number of bands	7	3
Data file size (pixels)	2944 x 3160	2972 x 3000

These two images were geometrically corrected with the *image-to-image* scheme. The Landsat image used in the first phase of the study was adopted to be used in the rectification of these new images. The Landsat image previously used was already rectified and geo-referenced to the UTM coordinate system. Thus, this image was used as the *master* while those new images (shown in Tab. 43) were used as *slaves*. Each new image was registered to the georeferenced image. These new data were then classified using the same procedures as those used in the first phase of the study. Landsat data were used with the land mask while SPOT XS data was used with the water mask. This was done because of striping pixels occurring on the Landsat image. Result from the classification is in the form of a raster image. To be able to integrate this data into a vector-based GIS used in the further step, the raster image had to be converted into a vector form.

For the physical criteria of the mangrove forest distribution, the altitude limitation and soil type referred above were adopted. To delineate the elevation 7.5 m, as the upper limit of the altitude factor, the interpolation was done using the surface model of this area. For the low tides of 0.6 m, the lower limit of the altitude factor, a special technique had to be implemented since the resulting DEM derived from the first part of the study contains *no data* in water areas (in the DEM). The Bathymetry for the offshore coastal zone was tried out to construct a surface model containing depth values. The resulting surface model can be used to interpolate depth curves in the same way as the elevations derived from a DEM or a TIN. This surface modeling was successfully performed both with the terrain analysis in the ERDAS system and with the TIN module of ARC/INFO.

To create a surface model with TIN for the water areas, available depth curves and depth points contained in hydrographic and topographic maps were extracted and digitized by means of ARC/INFO. Managing associated attribute tables were done. In this case, only depth point values were used in building a TIN. The depth values are soundings data measured by the Hydrographic Department. The coastlines was also used as a *hard break-line* in the process. This includes the coastline of two islands occurred in this study area. These islands were specified to be the *harderase* polygons to create holes in the TIN. This was done to influence the resulting triangulation in order to keep the interpolation not occurring inside the *erase* polygons. It is not necessary to rescale extracted z values in the case of building the TIN, because this can be done later to a resulting model whenever contours should be interpolated. As an alternative use, surface modeling for the water areas was also done using the methodology of constructing a DEM. Since terrain data, z values, should be stored in integer values, the z values were then multiplied by 10 during the process of building the DEM. In this way, interpolated contour lines with value such as 75 are identical with the 7.5 contour lines. However, the resulting DEM using this method is a grid-based form with resolution of 25 meters. This resolution is relatively coarse for this study case.

With the resulting TIN surface model, the depth curves such as 0.6 m can be now interpolated. Figure 90 illustrates the TIN surface model in the water of this study area. Figure 91 illustrates depth curves interpolated from the resulting surface model. However, it should be noted that this model was constructed using relatively out of date data. Up-to-date data are needed for building a good surface model for water areas. This is because seashores would be changed yearly according to the sedimentation.

Fig. 90: TIN surface model in the water area of the study area IV (Source: Author).

In short, altitudes of 2.5 and 7.5 m were interpolated from the surface model that has been created for inland of the study area., while the 0.6 depth was interpolated from the model of the water areas. All interpolated altitudes, i.e. 0.6, 2.5, and 7.5 m, were stored as the line feature in ARC/INFO *coverages*. These *coverages* were then combined into one *coverage*. For GIS overlay operations, this resulting coverage was then modified to include meaningful polygons. To do this, the boundary of the study area was combined in the coverage. Resulting polygons derived from the intersection between the boundary line and altitude lines were then labels according to the altitude ranges. Resulting coverage is illustrated in the Fig. 92. To perform spatial analysis operations, the attribute tables of both altitude coverage and soil coverage were performed. The site potential indices for mangrove forests based on the consideration of physical factors is shown in Tab. 44.

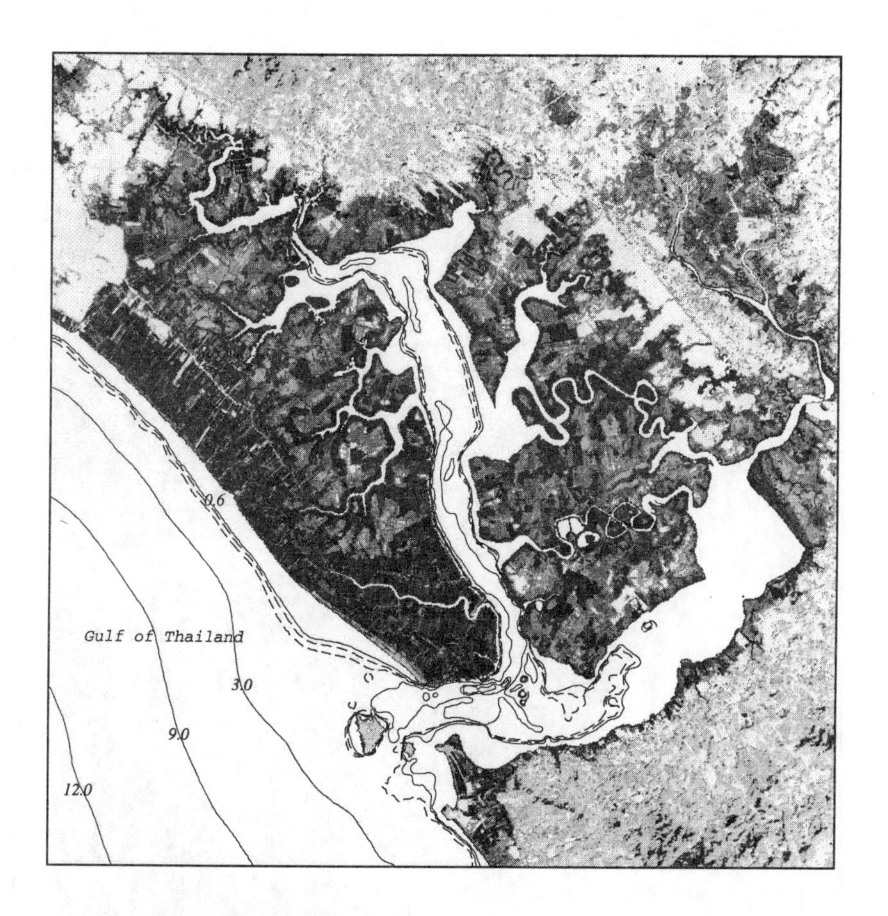

Fig. 91: **Depth curves interpolated from a surface model (Source: Author).**

Tab. 44: **Site potential for mangrove forests based on the consideration of physical factors.**

Site potentials	Altitude factor (m)	Soil factor
Suitable	Depth 0.6 - 2.5	Tha Chin series
Moderately suitable with altitude restrictions	2.5 - 7.5	Tha Chin series
Moderately suitable with soil restrictions	0.6 - 2.5	Cha-am series, Chonburi series, Bang Pakong series
Slightly suitable	2.5 - 7.5	Cha-am series, Chonburi series, Bang Pakong series
Could be suitable (needs more soil investigations)	0.6 - 2.5	Mud-flats
Not suitable (restricted to either altitude or soil)	lower 0.6 or above 7.5	Other soil types

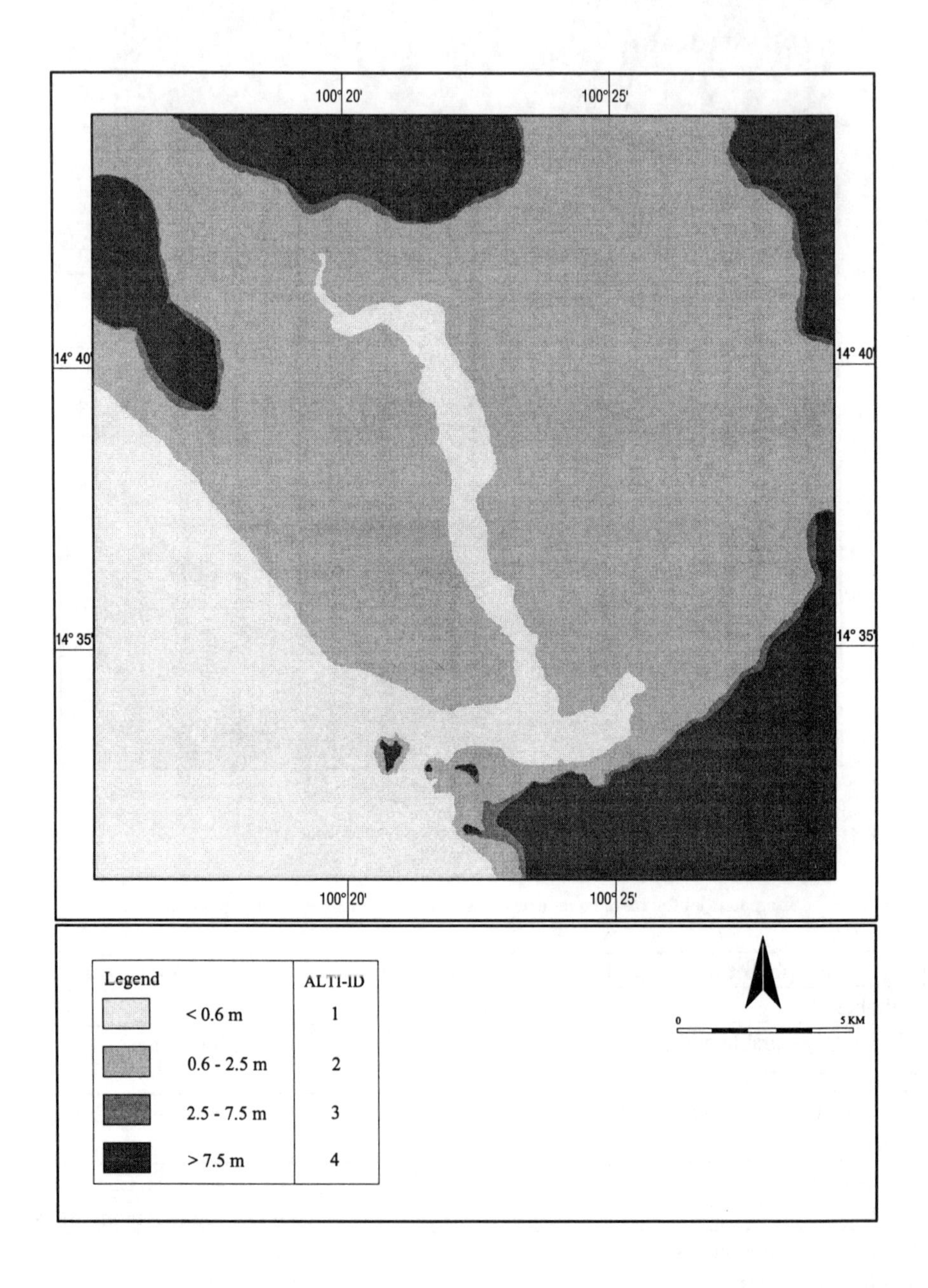

Fig. 92: Altitude ranges, criterion for site potential (Source: Author).

Before applying a GIS spatial operation, items- and features-ID in the attribute tables of all *coverages* were prepared for optimal use in the overlay operations. The relationship is summarized in Tab. 45. SOIL-ID numbers shown in this table are related to the description in the section 4.4 of this book and the soil map illustrated in Fig. 9.

Tab. 45: Relationship between site potential, criteria, and items in attribute tables (Author).

Site potential	INDEX	SOIL-ID	ALTI-ID
Suitable	1	4	2
Moderately suitable I	2	4	3
Moderately suitable II	3	5, 7, 12	2
Slightly suitable	4	5, 7, 12	3
Could be suitable	5	32	2
Not suitable	6	(others)	1 or 4

It has to be noted that the conservation zones measured from coastlines and streams had to be established. To do this, the extent of buffer zones along those streams were determined using the GIS buffering process. Due to the fact that resulting maps should be acceptable for practical use in land planing, the shorelines and streams used in this process are based on the details contained in topographic maps 1:50,000, not based on the coastlines delineated from satellite imagery. This is because the coastlines in topographic maps are accepted by the law. The shorelines and streams were included as *arcs* with different identification numbers. As part of the process by which the calculations for stream buffer characterization were made, it was necessary to set node points, wherever a transition from shorelines to streams occurred.

At this point, the entire set of data was ready for the GIS analysis. Dealing with narrow 20 m buffers required a spatial analysis operation by means of a vector-GIS instead of the grid-based GIS. By means of such a system the overlay of data layer has been performed. The zones of 100 m from coastlines and 20 m from streams were buffered. In addition to these zones, the boundaries of interesting areas were located and additionally defined 100 m from those buffers. That means, the spatial operation will be performed to analyze the study area within 3 zones, namely:

- Zone I, areas within 100 m from coastlines and 20 m from streams to the land
- Zone II, areas within 100 m from Zone I to the land, and
- Zone III, or water zone, areas from coastlines to the water areas.

Figure 93 shows the conceptual flowchart of GIS overlay operations applied in this stage and the commands series used for manipulating an attribute table to create a coverage containing potential sites for mangrove forests.

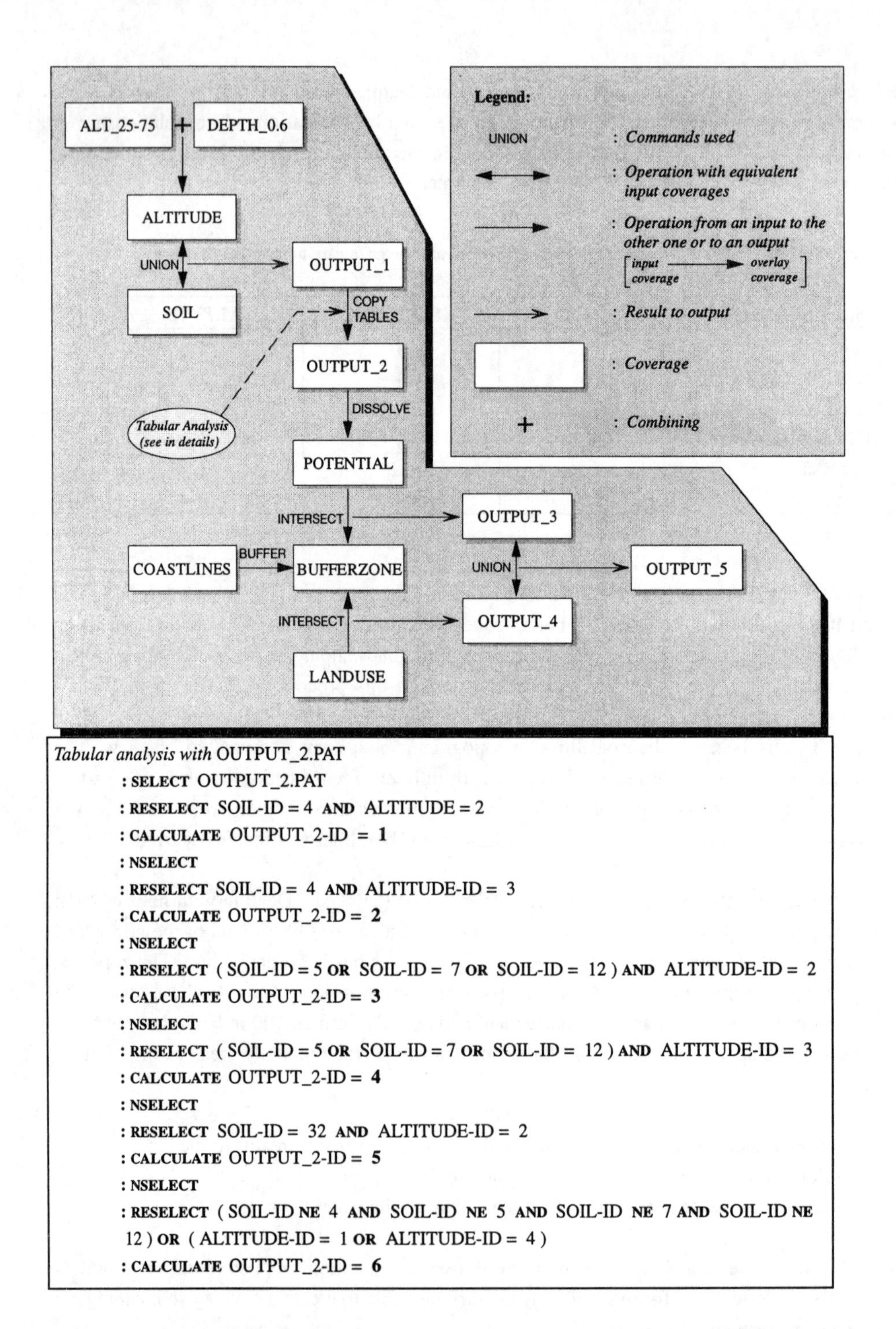

Fig. 93: Conceptual flowchart of GIS overlay operations applied in this case study (Source: Author).

When the spatial operation had been successfully performed, with UNION function to the ALTITUDE- and SOIL coverages, the derived coverage could be interpreted and used for mapping. According to Fig. 93, the 'POTENTIAL' *coverage* is the GIS operation output which can be interpreted for further use. Site potential index map for mangrove forest development in this coastal zone is derived from this *coverage*, as shown in Fig. 94. This thematic map could be beneficial for further management of the mangrove in the coastal zone. From field surveys, it was found that some parts of the coastal zone were composed of fresh sediments. New mangrove trees were found in these areas. This can be concluded that those areas are suitable for mangrove regeneration. This is a reason why the site potential "*Could be suitable* (in Tab. 45)" was included in the analysis.

According to the Fig. 93, two spatial operation steps were further done with INTERSECT and UNION commands to produce some more results of the analysis. Derived coverage, OUTPUT_5, contains also meaningful information in the form of spatial polygons and their attribute data. This coverage could be also beneficial for further use in managing and planning this coastal zone. A portion of the derived coverage was presented in the form of thematic map as shown in Fig. 95. An example of the query of information associated with polygons is then presented in Fig. 96. The results of these queries can be used to explain the areas, including the size-, the potential index-, the buffer zone-, and the existing land cover of the queried area. This information can be essential for decision makers in planning this coastal zone. For example, one can forecast where should be reforestation areas and which priority can be set, which type of mangrove should be adopted in such case, how many acres there are, and so far and so forth.

This case study indicates that the information of the GIS database can be retrieved, updated, added and analyzed at any time. The results of the analysis can be interpreted and presented in various forms, such as various types of thematic maps depending on the purposes, tables. These results could be beneficial for use in the area planning.

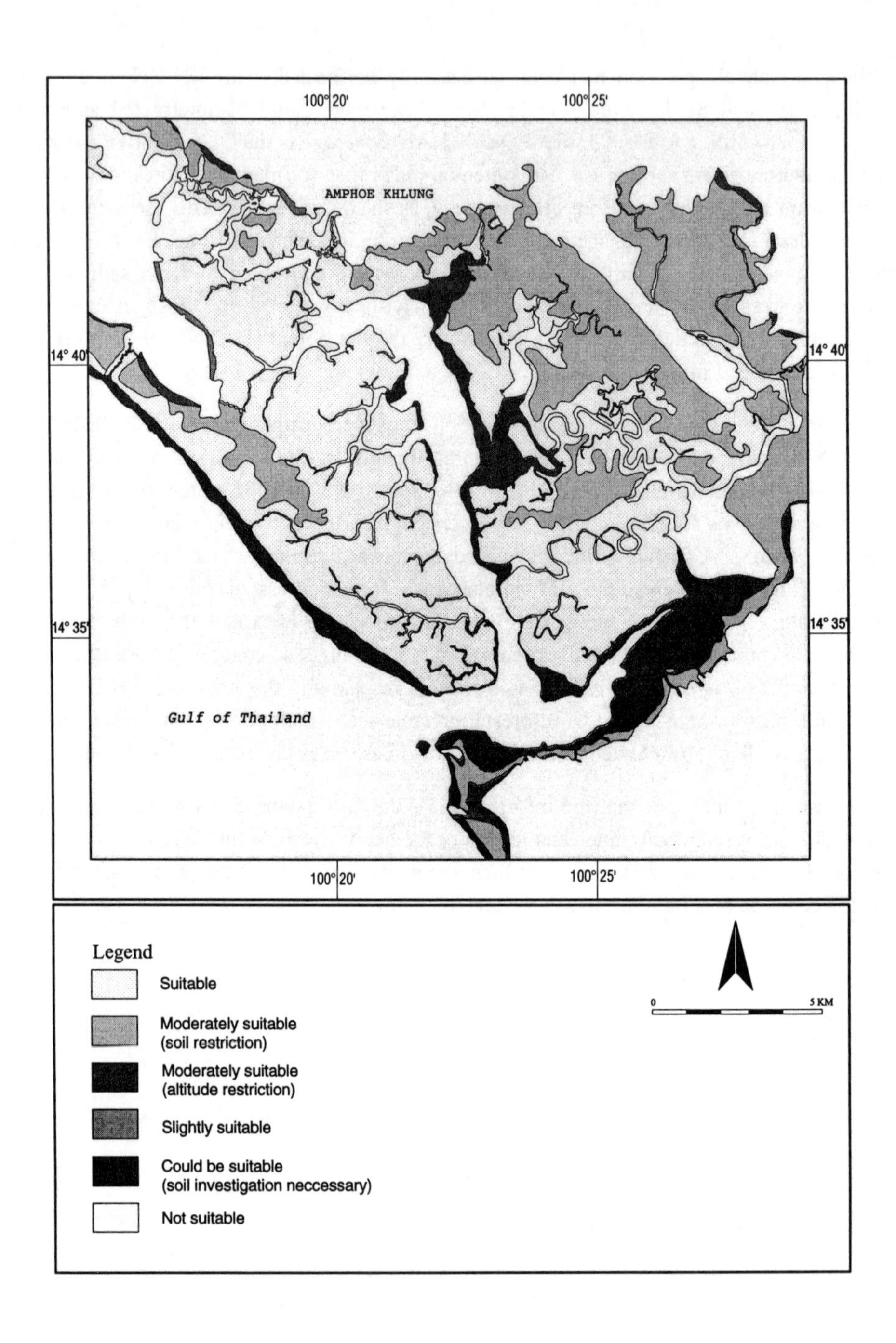

Fig. 94: Site potential index map for mangrove forest development of the study area IV (Source: Author).

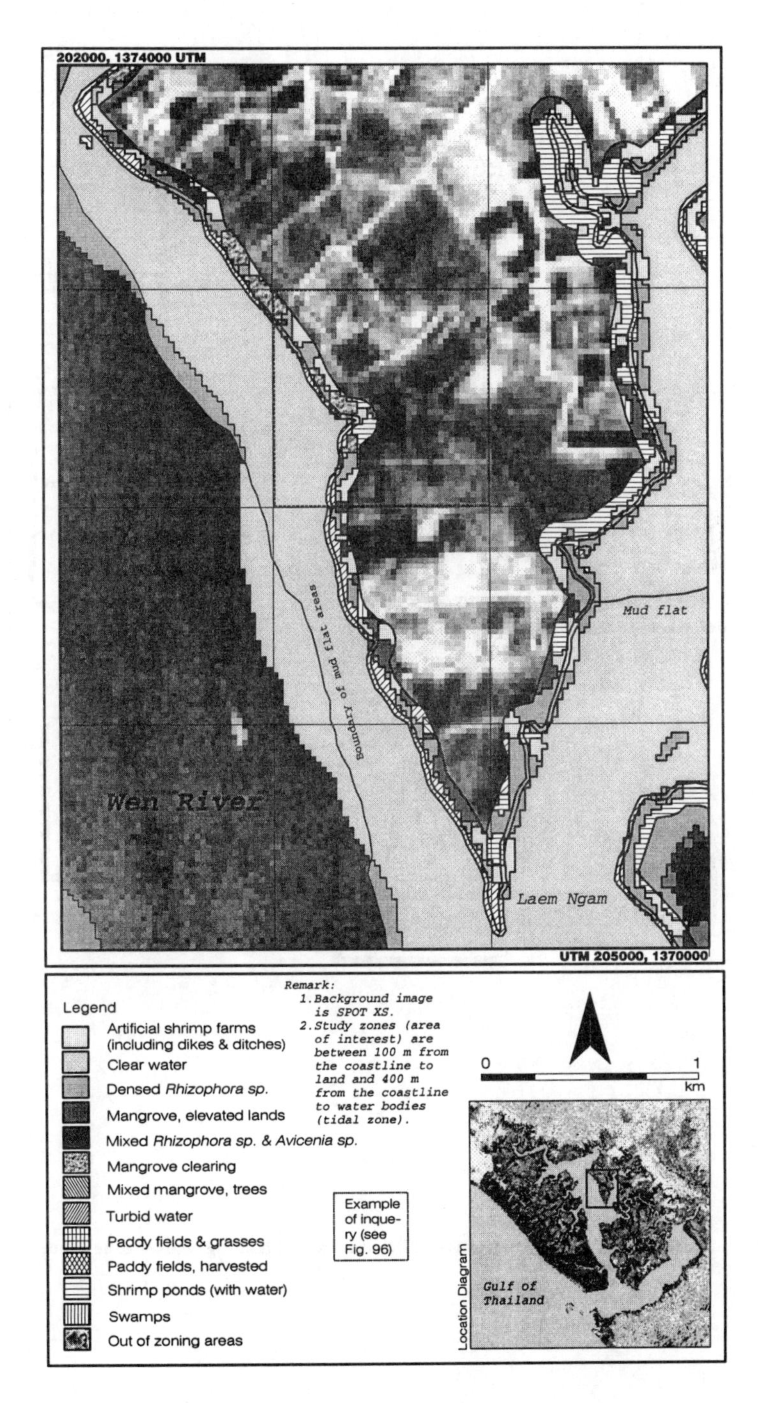

Fig. 95: **An example of the informational map needed for the coastal zone planning (a portion of the study area IV) (Source: Author).**

Geographisches Institut
der Universität Kiel

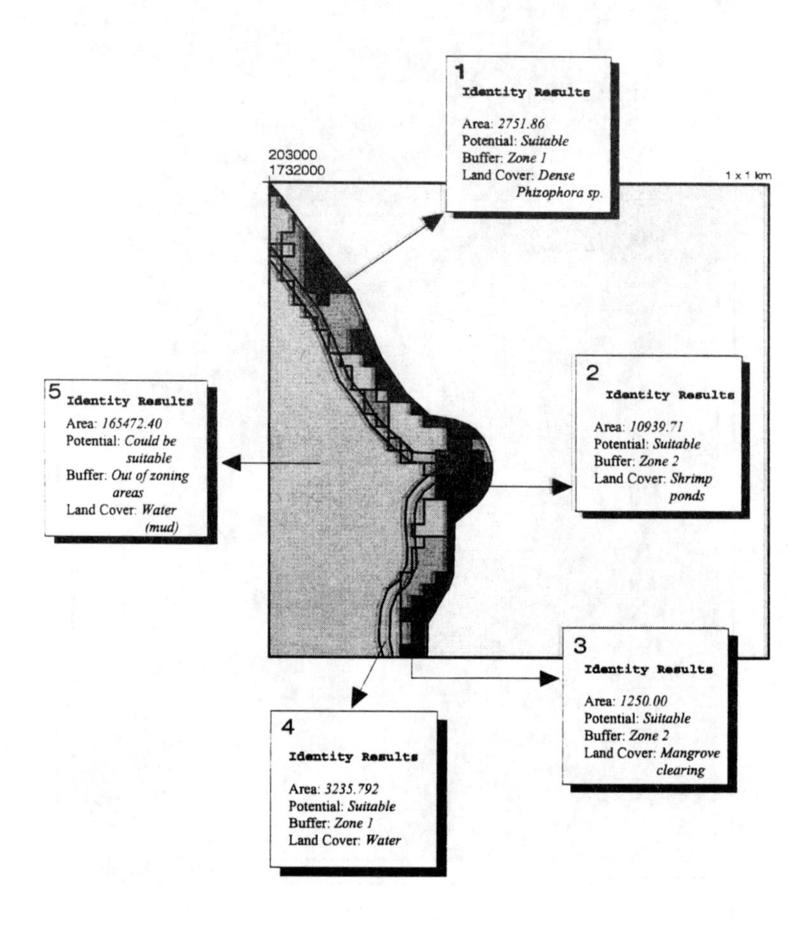

Fig. 96: **Example of the query from the GIS database of Khlung coastal zone (Source: Author).**

7.3 Discussion

The resulting land use/land cover classification was mapped in a scale of 1:50,000 (The map is shown with a reduced scale in Appendix E). It can be noted that the delineation of coastlines from imagery could lead to correct these baselines in topographic maps. Both active and idle shrimp ponds can be mapped. The principal vegetation categories such as grasslands, healthy or degraded mangroves can be identified. The occurrence and turbidity of water was successfully delineated. Swamp areas were also discriminated and should be included in mapping. This is because this land cover class could be an important information for land management in the future. It was found in this area that a large area of paddy fields have been abandoned. These areas looked like freshwater swamps covered with grasses and aquatic plants. Tidal-swamp areas which are subject to sea water was also found in this zone.

It is possible to construct an accurate inter-tidal terrain model (either DEM or TIN) using the data from the Hydrographic Department. In any cases, data from conventional surveys of the inter-tidal zone by geodetic levelling or sounding are preferable. These data are commonly accurate to about 1cm in height. Using the data already transferred to a map would cause more errors than the direct use of raw data. This is because the accuracy of measured depth data is subject to the scale of map used and the accuracy of the map itself. However, the propagation of errors would mostly happen for the planimetric (X and Y) coordinates of data points. The vertical accuracy would mostly be held systematically subject to the X-Y coordinates. Thus, the results reported in this case study may or may not be indicative of a terrain surface modeling for other areas. However, the comparison techniques in this study can be applied to other areas with different topographic profiles.

Using Landsat TM data incorporated into the spatial operation of a Geographic Information System is preferable for the study in the Region Level (see Fig. 30). For a specific area, this kind of data can be adopted with some limitations according to the resolution of the sensors. For example, the classified land use/land cover map using Landsat TM data is too coarse for the 20-meter buffer zone in our case. SPOT data would be preferable due to their better resolution.

Buffering coastlines delineated from satellite data must be carefully considered. The term 'coastlines' should be clearly definited due to the fact that coastlines contained in topographic maps are always accepted and protected by law.

As a summary of the chapter 7.2, it can be concluded that the integration of Remote Sensing and terrain surface model with Geographic Information System provides meaningful data. They could be beneficial for further studies relevant to the management of natural resources and environment in this zone. However, it must be noted that the accuracy of terrain models developed here were not verified. Normally, a vertical map accuracy can be defined as the RMS error in evaluation in terms of an evaluation of projection datum for well-defined points. The elevations of well-defined points determined from derived terrain data (or map in most cases) will be compared to corresponding elevations determined by a survey of higher accuracy for assessing the vertical accuracy. Thus, the results reported in this case study may or may not be applicable in practice, but the overall techniques and ideas of this chapter could be beneficial for other studies.

8 CONCLUSION

In Thailand, land use and land cover maps have been intensively used by the government agencies whose mission involves land and resource management planning. Some existing land use and land cover classification systems are based on aerial photographs with a high degree of details, while the others are based on visual satellite image interpretation with a coarse categorization. Since the use of high resolution satellite data is expected to provide a better understanding of major development trends in Thailand, this study tried a few key techniques. An attempt was shown to assess and modify the existing land use/land cover classification system with particular emphasis on the use of high resolution satellite imagery. These results may be a guide for use with computer-aided classification systems. The investigation of satellite image processing techniques as well as the integration of these data with surface models in the form of Digital Elevation Model (DEM) and Triangulated Irregular Network (TIN) was tried out. The Geographic Information system was adopted as a tool for supporting and managing the data of those systems.

Analyses of Landsat TM data for digital classification of land uses and land covers was carried out in four study areas located in various-environmental parts of Thailand. It indicates that the automated classification of Landsat TM data can yield results in high degree of details and can be useful in evaluating the distribution, quality, and change of natural resources, semi-natural resources, and man-activities on those lands. However, each of the techniques for satellite image processing used in the study has advantages for use in particular situations. Thus, it is not intended a judgement as to what the best technique might be in all cases.

Surface modeling can be performed both by means of DEM and TIN. The usefulness of the integration of these data to the satellite data classification of certain land use and land cover categories was investigated. When incorporated with Landsat TM multispectral data in a compatible form, the addition of the terrain information has been shown to improve classification accuracy. Normally, the construction of DEM and TIN can be successful using elevation data measured through standard surveying techniques or Global Positioning Systems (GPS). The analysis given here provides a simple technique for creating DEM and TIN from available topographic maps. The results obtained from DEM and TIN data are interesting and revealing, because they can be used to derive a wealth of information of a land surface. These surface models are very useful for geographic data reduction, geomorphometrical interpretation, and various manipulations within a geographic information system. The accuracy of these surface model data is dependent on the quality of the collected data and the subsequent interpolation algorithms. It has been determined that the accuracy and detail of geographic data that can be automatically extracted from either DEM or TIN are directly related to the quality and resolution of the DEM or TIN itself. The possibility of DEM generation from topographic maps seems to be currently limited by the computer configurations commonly employed for image processing and mapping tasks. However, elevation data stored in the digital form such a DEM appear to be a convenient and cost-effective means of acquiring terrain data for input into a Geographic Information System. This data should be available in Thailand.

Based on the results of the study, it can be concluded that the use of remotely sensed data, terrain surface models and GIS procedures provide the successfulness of developing land use/land cover classification system. The developed system and the methodology applied in this study could provide potentially powerful inventory and management tools for land planning in various ways and purposes. In order that the methods and ideas of this study might be applied on a wide scale, setting working groups for considering the possibility is recommended. The working groups should consist of the representative from government agencies and academic institutions whose works and interests are related to remote sensing and GIS. The groups should specific an optimal land use/land cover system that is best suited as a standard one for world-wide use without restriction. Researchers should not have to develop an own one every time they will do a research. The results of this study in any aspects, in particular about the developed land use/land cover classification system, crop calendars, information charts, the methods of surface modeling, the methods of analyzing land use/land cover categories with various factors and the methods of GIS database establishment could be support the consideration of the working group.

If possible, land use/land cover maps in the scale of 1:50,000 for the whole country should be produced using high resolution remotely sensed data, as prepared in this study (Unfortunately, the maps attached with this book were reduced the scale). Resulting thematic maps could be adopted to further use in various cases. This is important because the land use/land cover map of the same area and time should be contained the same information both in qualitative and quantitative meanings.

In addition, the development of GIS database concerning land management should be established under the corporation with both government and non-government agencies. Available data from various sources should also be firstly take into account. The duration and period of time that the GIS database should be updated should also be considered. Since the computer-assisted system should be adopted for this task and due to the fact that the technologies of computer system have rapidly changed, an optimal system being established should be considerably selected to be installed at institutes concerned.

9 ABSTRACT

An attempt to evaluate the usefulness of Landsat TM data for automated classification of land use and land cover in Thailand is conducted. Developing a land use/land cover classification system is carried out in order to solve the problem of the incompatibility among existing classification systems derived from various sources. This system is intended to be ease used with data derived from high resolution satellites such as Landsat TM by means of computer-assisted classification. In addition to the operational procedures, the surface modeling performed both by means of DEM and TIN concepts is implemented to this effort throughout the study in many ways. Four study areas located in various-environmental parts of Thailand are selected to test in this study. Designated land use/land cover classification system, significant analysis techniques, and relevant information in forms of tables, charts, thematic maps, etc., are presented. Finally, the usefulness of the resulting data and techniques has been further demonstrated on the basis of possibility and suitability in a study area in the coastal zone. The results from this test could be essential for land use management planning in this coastal zone.

This research points out that standardizing land use/land cover classification systems is urgently important. Working groups from institutes concerned should be performed to deal with this task. Furthermore, producing a standard form of land use/land cover maps scale to 1:50,000 as well as elevation data in digital formats for the whole country is also recommended, since these data are basically needed for any further studies.

10 ZUSAMMENFASSUNG

Behörden und Institute in Thailand arbeiten derzeit mit jeweils eigenen Landnutzungsklassen und Erfassungsmethoden. Diese basieren auf unterschiedlichen Sourcen und sind nicht kompatibel. Die Entwicklung von Standards zum Datenaustausch wird angestrebt. Ziel dieser Arbeit ist es, ein allgemeingültiges System zur Landnutzungsklassifikation ganz Thailands zu entwickeln. Dabei werden die Möglichkeiten der Integration von hochauflösenden Satellitendaten in Betracht gezogen. Eine Bewertung der Eignung von Landsat TM-Daten wird durchgeführt, unterstützt durch Geländemodellierungen auf der Basis von DGM- und TIN-Konzepten. An vier extrem unterschiedlichen Testgebieten wird das Verfahren auf die Naturraumvielfalt Thailands abgestimmt. Die Klassifikationseinheiten, das Analyseverfahren, die Datenbank und die resultierenden Karten werden eingehend beschrieben. Schließlich wird die Anwendbarkeit des Systems für weiterführende Studien exemplarisch aufgezeigt, indem für ein Testgebiet Nutzungsmöglichkeiten für ein Küstenzonenmanagement erarbeitet werden.

Die Ergebnisse zeigen deutlich, daß ein Standard der Landnutzungsklassifikation möglich ist. Es wird empfohlen, das entwickelte Klassifikationssystem instituts- und behördenübergreifend anzuwenden und zu prüfen, einheitliche Landnutzungskarten im Maßstab 1:50000 und einheitliche digitale Geländemodelle für ganz Thailand zu erstellen.

11 BIBLIOGRAPHY

AHERN, F. J. & J. SIROIS (1989): Reflectance Enhancements for the Thematic Mapper An Efficient Way to Produce Images of Consistently High Quality. *In:* Photogrammetric Engineering and Remote Sensing, 55(1): 61-67.

ALBERTZ, J. & A. MEHLBREUER (1989): Production of Satellite Image Maps from TM Data. *In:* Proceedings of a Workshop on 'Earthnet Pilot Project on Landsat Thematic Mapper Applications', held at Frascati, Italy in December 1987: 315-321.

ANDERSON, J. R., E. E. HARDY, J. T. ROACH & R. E. WITMER (1976): A Land Use and Land Cover Classification for Use with Remote Sensor Data. US. Geological Survey Professional Paper. 964.

ANUTA, P., L. BARTILUCCI, E. DEAN, F. LOZANO, E. MALARET, C. McGILLEM, J. VALDES & C. VALENZUELA (1984): Landsat-4 MSS and Thematic Mapper Data Quality and Information Content Analysis. *In:* Proceedings of IGARSS, 84 Symposium, Strasbourg. 27-30 August 1984: 85-92.

ARONOFF, S. (1982): Classification Accuracy : A User Approach. *In:* Photogrammetric Engineering and Remote Sensing, 48(8): 1299-1307.

ARONOFF, S. (1985): The Minimum Accuracy Value as an Index of Classification Accuracy. *In:* Photogrammetric Engineering and Remote Sensing, 51(1): 99-111.

AUNG, Z. (1991): The Study of Landslide Susceptibility Using the GIS Approach (West of Amphoe Phi Pun, Nakhon Si Thammarat Province). Master Thesis, Asian Institute of Technology (AIT).

BANNINGER, C. (1987): Discrimination of Geobotanical Anomalies in Coniferous Forests from Landsat TM Data. *In:* Proceedings of a Workshop on 'Earthnet Pilot Project on Landsat Thematic Mapper Applications', held at Frascati, Italy in December 1987: 233-241.

BELWARD, A. S. & J. C. TAYLOR (1986): The Influence of Resampling Method and Multitemporal Landsat Imagery on Crop Classification Accuracy in The United Kingdom. *In:* Proceedings of IGARSS, 86 Symposium, Zurich, 8-11 September 1986: 529-535.

BELWARD, A. S., J. C. TAYLOR, M. J. STUTTARD, E. BIGNEL, J. MATHEWS & D. CURTIS (1990): An Unsupervised Approach to the Classification of Semi- Natural Vegetation from Landsat Thematic Mapper Data. *In:* International Journal of Remote Sensing, 11(3): 429-445.

BENSON, A. S. & S. D. DeGLORIA (1985): Interpretation of Landsat-4 Thematic Mapper and Multispectral Scanner Data for Forest Surveys. *In:* Photogrammetric Engineering and Remote Sensing, 51(9): 1281-1289.

BOLSTAD, P. V. & T. M. LILLESAND (1992): Rule-Based Classification Models: Flexible Integration of Satellite Imagery and Thematic Spatial Data. *In:* Photogrammetric Engineering and Remote Sensing, 58(7): 965-971.

BOLSTAD, P. V. & T. STOWE (1994): An Evaluation of DEM Accuracy: Elevation, Slope, and Aspect. *In:* Photogrammetric Engineering and Remote Sensing, 60(11): 1327-1332.

BOONYOBHAS, C. (1988): Forest Inventory in Thailand. Paper presented at RASD/86/049 Steering Committee Meeting, Bangkok, 25-26 January 1988. *In:* Strengthening Forestry Inventory Capabilities for Forest Management in Asia and the Pacific: 112-139.

BOOTH, D. J. & R. B. OLDFIELD (1989): A Comparison of Classification Algorithms in terms of Speed and Accuracy after the Application of a Post-Classification Model Filter. *In:* International Journal of Remote Sensing, 10(7): 1271-1276.

BURROUGH, P. A. (1986): Principle of Geographic Information Systems for Land Resources Assessment. Oxford. 194.

CALOZ, R. & T. BLASER (1987): Large-Scale Forest Management Using Landsat Thematic Mapper Data. *In:* Proceedings of a Workshop on 'Earthnet Pilot Project on Landsat Thematic Mapper Applications', held at Frascati, Italy in December 1987: 287-291.

CAMPBELL, J. B. (1981): Spatial Correlation Effects upon Accuracy of Supervised Classification of Land Cover. *In:* Photogrammetric Engineering and Remote Sensing, 47(3): 355-363.

CANAS, A. A. D. & M. E. BARNETT (1985): The Generation and Interpretation of False-Colour Composite Principal Component Images. *In:* International Journal of Remote Sensing, 6(6): 867-881.

CARTER, J. R. (1988): Digital Representation of Topographic Surfaces. *In:* Photogrammetric Engineering and Remote Sensing, 54:(11): 1577-1580.

CENTER OF INFORMATION SYSTEM (1994): Geographic Information System Index. Center of Information System, Office of The Permanent Secretary of Ministry of Science and Technology. Bangkok. 351.

CHAVEZ, P. S. (1989): Extracting Spectral Contrast in Landsat Thematic Mapper Image Data Using Selective Principal Component Analysis. *In:* Photogrammetric Engineering and Remote Sensing, 55(3): 339-348.

CHANGJATURAS, S. (1989): Data Storage Based on Geographic Information System for Land Evaluation of Doi Thung Watershed, Changwat Chiang Rai. Master Thesis, Kasetsart University, Bangkok, Thailand.

CHUVIECO, E. (1987): Multitemporal Analysis of TM Images : Application to Forest Fire Mapping and Inventory in a Mediterranean Environment. *In:* Proceedings of a Workshop on

'Earthnet Pilot Project on Landsat Thematic Mapper Applications', held at Frascati, Italy in December 1987: 279-285.

CHUVIECO, E. & R. G. CONGALTON (1988): Using Cluster Analysis to improve the Selection of Training Statistics in Classifying Remotely Sensed Data. *In:* Photogrammetric Engineering and Remote Sensing, 54(9): 1275-1281.

CIBULA, G. W. & M. O. NYQUIST (1987): Use of Topographic and Climatological Models in a Geographical Data Base to Improve Landsat MSS Classification for Olympic National Park. *In:* Photogrammetric Engineering and Remote Sensing, 53(1): 67-75.

CIVCO, D.L. (1989): Topographic Normalization of Landsat Thematic Mapper Digital Imagery. *In:* Photogrammetric Engineering and Remote Sensing, 55(9): 1303-1309.

CLARK, W.C. (1990): Managing Planet Earth. *In:* Scientific American Special Issue: Managing Planet Earth, 261(3): 18-28.

CONESE, C., G. MARACCHI, F. MIGLIETTA & F. MASELLI (1988): Forest Classification by Principal Component Analysis of TM Data. *In:* International Journal of Remote Sensing, 9(10): 1897-1612.

CONGALTON, R. G. (1988): A Comparison of Sampling Schemes Used in Generating Error Matrices for Assessing the Accuracy of Maps Generated from Remotely Sensed Data. *In:* Photogrammetric Engineering and Remote Sensing, 54(5): 593-600.

CONGALTON, R. G. (1988): Using Spatial Auto-Correlation Analysis to Explore the Errors in Maps Generated from Remotely Sensed Data. *In:* Photogrammetric Engineering and Remote Sensing, 54(5): 587-592.

CONGALTON, R. G. & R. A. MEAD (1983): A Quantitative Method to Test for Consistency and Correctness in Photo-interpretation,. *In:* Photogrammetric Engineering and Remote Sensing, 49(1): 69-74.

CONGALTON, R. G., R. G. ODERWALD & R. A. MEAD (1983): Assessing Landsat Classification Accuracy Using Discrete Multivariate Analysis Statistical Techniques. *In:* Photogrammetric Engineering and Remote Sensing, 49(12): 1671-1678.

CRIST, E. P. & R. C. CICONE (1984): Application of the Tasselled Cap Concept to Simulated Thematic Mapper Data. *In:* Photogrammetric Engineering and Remote Sensing, 50(3): 343-352.

CRIST, E. P. & R. J. KAUTH (1986): The Tasselled Cap De-Mystified. *In:* Photogrammetric Engineering and Remote Sensing, 52(1): 81-86.

DEPARTMENT OF LAND DEVELOPMENT (1981): Land Use Plan of Khon Kaen Province. Department of Land Development, Bangkok, Thailand (version Thai). 196.

DEPARTMENT OF LAND DEVELOPMENT (1983a): Land Use Plan of Chantaburi Province. Department of Land Development, Bangkok, Thailand (version Thai). 213.

DEPARTMENT OF LAND DEVELOPMENT (1983b): Land Use Plan of Trat Province. Department of Land Development, Bangkok, Thailand (version Thai). 181.

DEPARTMENT OF LAND DEVELOPMENT (1987): Land Use Plan of Lampang Province. Department of Land Development, Bangkok, Thailand (version Thai). 226.

DEPARTMENT OF LAND DEVELOPMENT (1990): Land Use Plan of Angthong Province. Department of Land Development, Bangkok, Thailand (version Thai). 75.

DITBANJONG, D. (1990): Soil Erosion Mapping in the *Dok Rai* Reservoir area of Rayong Province Using Remote Sensing and GIS (version Thai). *In:* Proceedings of the Seminar on Remote Sensing and GIS for Soil and Water Management, Khon Kaen, December 1990: 216-223.

ERDAS (1991a): ERDAS Field Guide. Second Edition. ERDAS Inc., Atlanta. 394.

ERDAS (1991b): Image Processing and Multivariate Image Analysis. ERDAS Inc., Atlanta. 282.

ERDIN, K., A. HIZAL & E. ATAMAN (1986): Landsat-5 TM Data Applications to Land Use Classification on around the Bosphorus Area, Turkey. *In:* Proceedings of IGARSS, 86 Symposium, Zurich, 8-11 September 1986: 1143-1148.

ESRI (1992a): ARC/INFO Data Model, Concepts, & Key Terms. Environmental System Research Institute, Inc.

ESRI (1992b): ARC/INFO Surface Modeling with TIN™. Environmental System Research Institute, Inc.

ESTES, J. E., D. STOW & J. R. JENSEN (1982): Monitoring Land Use and land Cover Changes, *In:* Remote Sensing for Resource Management. Soil Conservation Society of America, Iowa: 100-110.

FAO/UNEP (1995): Background Note On-going Activities Relating to Land Use and Land Cover Classification. *In:* Initiative on Standardisation of Land Use and Land Cover Classification Systems. Circular letter No. 4, August 1995.

FRIZPATRIK-LINS, K. (1981): Comparison of Sampling Procedures and Data Analysis for a Land Use and Land Cover Map. *In:* Photogrammetric Engineering and Remote Sensing, 47(3): 343-351.

FRANK, D. T. (1988): Mapping Dominant Vegetation Communities in the Colorado Rocky Mountain Front Range with Landsat Thematic Mapper and Digital Terrain Data. *In:* Photogrammetric Engineering and Remote Sensing, 54(12): 1727-1734.

FRANKLIN, S. E. (1994): Discrimination of Subalpine Forest species and Canopy Density Using Digital CASI, SPOT PLA, and Landsat TM Data. *In:* Photogrammetric Engineering and Remote Sensing, 60(10): 1233-1241.

GONG, P. & P. J. HOWARTH (1990): An Assessment of Some Factors Influencing Multispectral Land Cover Classification. *In:* Photogrammetric Engineering and Remote Sensing, 56(5): 597-603.

HÄME, T. (1984): Landsat-Aided Forest Site Type Mapping. *In:* Photogrammetric Engineering and Remote Sensing, 50(8): 1175-1183.

HASTINGS, P. C. BOONRAKSA, A. RESANOND, S. CHAYAWATANAKIJJA & J. PANTHI (1991): Integrated Information for Natural Resources Management. Thailand Research Institute Foundation, Bangkok, Thailand. 144.

HOPKINS, P. F., A. L. MACLEAN & T. M. LILLESAND (1988): Assessment of Thematic Mapper Imagery for Forestry Applications under Lake States Conditions. *In:* Photogrammetric Engineering and Remote Sensing, 54(1): 61-68.

HUTCHINSON, C. F. (1982): Techniques for Combing Landsat and Ancillary Data for Digital Classification Improvement. *In:* Photogrammetric Engineering and Remote Sensing, 48(1): 123-130.

INGEBRITSEN, S. E. & R. J. P. LYON (1985): Principal Component Analysis of Multitemporal Image Pairs. *In:* International Journal of Remote Sensing, 6(5): 687-696.

ISAACSON, D. L. & W. J. RIPPLE (1990): Comparison of 7.5-Minute and 1-Degree Digital Elevation Models. *In:* Photogrammetric Engineering and Remote Sensing, 56(11): 1523-1527.

JAPAN INTERNATIONAL COOPERATION AGENCY (JICA) (1988): Aerial Photography and Forest Management Plan in the Encroached National Reserve Forest in the Kingdom of Thailand. Japan International Cooperation Agency (JICA), Tokyo, Japan: 151.

JANSSEN, L. L. F. & F. J. M. van der WEL (1994): Accuracy Assessment of satellite Derived Land-Cover data: A Review. *In:* Photogrammetric Engineering and Remote Sensing, 60(4): 419-426.

JENSEN, J. R. (1986): Introductory Digital Image Processing. Prentice-Hall, New Jersey. 379.

JIAJU, L. (1988): Development of Principal Component Analysis Applied to Multitemporal TM Data. *In:* International Journal of Remote Sensing, 9(12): 1895-1907.

JUSTICE, C. O. & J. R. G. TOWNSHEND (1981): The Use of Landsat Data for Land Cover Inventories of Mediterranean Lands, *In:* Terrain Analysis and Remote Sensing: 133-153.

KADRO, A. (1987): Use of Land TM Data for Forest Damage Inventory. *In:* Proceedings of a Workshop on 'Earthnet Pilot Project on Landsat Thematic Mapper Applications', held at Frascati, Italy in December 1987: 261-270.

KARTERIS, M. A. (1990): The Utilities of Digital Thematic Mapper Data for Natural Resources Classification. *In:* International Journal of Remote Sensing, 11(9): 1589-1598.

KHON KAEN UNIVERSITY (1987): Ecosystems Interactions in A Rural Landscape: The case of Phu Wiang Watershed, Northeast Thailand. Khon Kaen University, Thailand: 172.

LABOVITZ, M. L. (1986): Issues Arising from Sampling Designs and Band Selection in Discriminating Ground Reference Attributes Using Remotely Sensed Data. *In:* Photogrammetric Engineering and Remote Sensing, 52(2): 201-211.

LATHROP, R. G. Jr. & T. M. LILLESAND (1986): Use of Thematic Mapper Data to Assess Water Quality in Green Bay and Central Lake Michigan. *In:* Photogrammetric Engineering and Remote Sensing, 52(5): 671-680.

LAUVER C. L. (1993): A Hierarchical Classification of Landsat TM Imagery to Identify Natural Grassland Areas and Rare Species Habitat. *In:* Photogrammetric Engineering and Remote Sensing, 59(5): 627-634.

LEPRIEUR, C. E. & J. M. DURAND (1988): Influence pf Topography on Forest Reflectance Using Landsat Thematic Mapper and Digital Terrain Data, 54(4): 491-496.

LILLESAND, T. M. & R. W. KIEFER (1994): Remote Sensing and Image Interpretation, Third Edition. John Wiley, New York: 750.

LO, C. P. (1986): Applied Remote Sensing. Longman Scientific and Technical: 393.

LOELKES, G. L., G. E. HOWARD, E. L. SCHWERTZ, P. D. LAMPERT & S. W. MILLER (1983): Land Use/Land Cover and Environmental Photointerpretation Keys. U.S. Geological Survey Bulletin 1600. 142.

MAJOR, D. J., F. BARET & G. GUYOT (1990) A Ratio Vegetation Index Adjusted for Soil Brightness. *In:* International Journal of Remote Sensing, 11(5): 727-740.

MAUSEL, P. W., W. J. KRAMBER & J. K. LEE (1990): Optimum Band Selection for Supervised Classification of Multispectral Data. *In:* Photogrammetric Engineering and Remote Sensing, 56(1): 55-60.

MAUSER, W. (1989): Agricultural Land Use Classification in the Upper Rhine Valley Using Multitemporal TM Data. *In:* Proceedings of a Workshop on 'Earthnet Pilot Project on Landsat Thematic Mapper Applications', held at Frascati, Italy in December 1987: 191-198.

MEAD, R. A. & J. SZAJGIN (1982): Landsat Classification Accuracy Assessment Procedures. *In*: Photogrammetric Engineering and Remote Sensing, 48(1): 139-141.

MIDDELKOOP, H. & L. L. F. JANSSEN (1991): Implementation of Temporal Relationships in Knowledge Based Classification of Satellite Images, Photogrammetric Engineering and Remote Sensing, 57(7): 937-945.

NATIONAL RESEARCH COUNCIL OF THAILAND (1990): TM CCT Format. Thailand Remote Sensing Center, National Research Council of Thailand, Bangkok.

NAUGLE, B. I. & J. D. LASHLEE (1992): Alleviating Topographic Influences on Land Cover Classification for Mobility and Combat Modelling. *In:* Photogrammetric Engineering and Remote Sensing, 58(8): 1217-1221.

OFFICE OF AGRICULTURAL ECONOMICS (1982): Crop Calendars and the Statistics of Period of Products Distribution from Farmers' accounts for the Crop Year 1980/81. Office of Agricultural Economics, Bangkok, Thailand.

OFFICE OF AGRICULTURAL ECONOMICS (1989): Agricultural Statistics of Thailand Crop Year 1988/89. Office of Agricultural Economics, Bangkok, Thailand.

OFFICE OF AGRICULTURAL ECONOMICS (1992): Agricultural Resource Mapping Using Satellite Data for the Northern, Northeast, and Eastern of Thailand. Office of Agricultural Economics, Bangkok, Thailand. 137.

OLSSEN, L. (1986): Symposium Remote Sensing for Resources Development and Environment Management, Enshede, August 1986.

OMAKUPT, M. (1978): Land Use Inventory of North Thailand Using Landsat Imagery. *In:* Proceedings of The Twelfth International Symposium on Remote Sensing of Environment. Environmental Research Institute of Michigan, Ann Arbor, Michigan: 2297-2306.

ONGSOMWANG, S. (1993): Forest Inventory, Remote Sensing and GIS for Forest management in Thailand. Berliner geographische Studien Band 38, Berlin, Germany. 272.

PRISLEY, S. P. & J. L. SMITH (1987): Using Classification Error Matrices to Improve the Accuracy of Weighted Land Cover Models. *In:* Photogrammetric Engineering and Remote Sensing, 53(9): 1259-1263.

REMOTE SENSING DIVISION (1991): Remote Sensing and Mangroves Project (Thailand). National Research Council of Thailand (NRCT). Funny Publishing Partnership, Bangkok, Thailand. 183.

RITTER, P. (1987): A Vector-Based Slope and Aspect Generation Algorithm. *In:* Photogrammetric Engineering and Remote Sensing, 53(8): 1109-1111.

ROBINSON, A. H., J. L. MORRISON, P. C. MUEHRCKE, A. J. KIMERLING & S. C. GUPTILL (1995). Elements of Cartography, Sixth Edition. John Wiley & Sons, Inc. 674.

ROSENFIELD, G. H. (1982): Sample Design for Estimating Change in Land Use and Land Cover. *In:* Photogrammetric Engineering and Remote Sensing, 48(5): 793-801.

ROSENFIELD, G. H. & K. FITZTRICK-LINS (1986): A Coefficient of Agreement as a Measure of Thematic Classification Accuracy. *In:* Photogrammetric Engineering and Remote Sensing, 52(2): 223-227.

SANGUANPONG, C. (SM.I) (1993): Physical and Biological Database Management of Sakaerat Environment Research Station (SERS). Master Thesis, Kasetsart University, Bangkok, Thailand.

SATTERWHITE, M., W. RICE & J. SHIPMAN (1984): Using Landform and Vegetative Factors to Improve the Interpretation of Landsat Imagery. *In:* Photogrammetric Engineering and Remote Sensing, 50(1): 83-91.

SCHARDT, M. (1987): Forest Classification with TM Data in the Area of Freiburg, Federal Republic of Germany. *In:* Proceedings of a Workshop on 'Earthnet Pilot Project on Landsat Thematic Mapper Applications', held at Frascati, Italy in December 1987: 251-259.

SHEFFIELD, C. (1985): Selecting Band Combinations from Multispectral Data, Photogrammetric Engineering and Remote Sensing, 51(1): 681-687.

SILAPATHONG, C. (1992): Utilisation Combinee D'un Systeme D'information Geographique et de la Tèlèdètection pour le Suivi et L'Amenenagement des Mangroves en Thailande. Thèse du Doctorat d'Ecologie Tèlèdètection Spatiale, Univ. Paul Sabatier (Toulouse III). 184.

SINGH, A. & A. HARRISON (1985): Standardised Principal Components. *In:* International Journal of Remote Sensing, 6(6): 883-896.

SKIDMORE, A. K. (1989): Unsupervised Training Area Selection in Forests Using a Nonparametric Distance Measure and Spatial Information. *In:* International Journal of Remote Sensing, 10(1): 133-146.

SKIDMORE, A. K. & B. J. TURNER (1988): Forest Mapping Accuracies Are Improved Using a Supervised Nonparametric Classifier with SPOT Data. *In:* Photogrammetric Engineering and Remote Sensing, 54(10): 1415-1421.

SKIDMORE, A. K. & B. J. TURNER (1992): Map Accuracy Assessment Using Line Intersect Sampling. *In:* Photogrammetric Engineering and Remote Sensing, 58(10): 1453-1457.

SMITINAND, T., S. SABHASRI & P. KUNSTADTER (1978): The Environment of Northern Thailand. *In:* KUNSTADTER, P., E. C. CHAMPION and S. SABHASRI (Ed.): Farmers in Forest. Honolulu: 24-40.

SOIL SURVEY DIVISION (1973): Report on Soil Survey of Khon Kaen Province, Detailed Reconnaissance Soil Map scale to 1:100,000. Soil Survey Division, Department of Land Development, Bangkok, Thailand.

SOIL SURVEY DIVISION (1976): Report on Soil Survey of Angthong Province, Detailed Reconnaissance Soil Map scale to 1:100,000. Soil Survey Division, Department of Land Development, Bangkok, Thailand.

SOIL SURVEY DIVISION (1977a): Report on Soil Survey of Chantaburi Province, Detailed Reconnaissance Soil Map scale to 1:100,000. Soil Survey Division, Department of Land Development, Bangkok, Thailand.

SOIL SURVEY DIVISION (1977b): Report on Soil Survey of Trat Province, Detailed Reconnaissance Soil Map scale to 1:100,000. Soil Survey Division, Department of Land Development, Bangkok, Thailand.

SOIL SURVEY DIVISION (1981): Report on Soil Survey of Lampang Province, Detailed Reconnaissance Soil Map scale to 1:100,000. Soil Survey Division, Department of Land Development, Bangkok, Thailand.

SPJELKAVIK, S. & A. ELVEBAKK (1989): Mapping Winter Grazing Areas for Reindeer on Svalbard Using Landsat Thematic Mapper Data. *In:* Proceedings of a Workshop on 'Earthnet Pilot Project on Landsat Thematic Mapper Applications', held at Frascati, Italy in December 1987: 199-205.

STEHMAN, S. V. (1992): Comparison of Systematic and Random Sampling for Estimating the Accuracy of Maps Generated from Remotely Sensed Data. *In:* Photogrammetric Engineering and Remote Sensing, 58(9): 1343-1350.

STENBACK, J. M. & R. G. CONGALTON (1990): Using Thematic Mapper Imagery to Examine Forest Understory. *In:* Photogrammetric Engineering and Remote Sensing, 56(9): 1285-1290.

STORY, M. & R. G. CONGALTON (1986): Accuracy Assessment : A User Perspective. *In:* Photogrammetric Engineering and Remote Sensing, 52(3): 397-399.

STRAHLER, A. H., T. L. LOGAN & N. A. BRAYANT (1978): Improving Forest Cover Classification Accuracy from Landsat by Incorpolating Topographical Information. *In*: Proc. 12th Int. Symp. on Remote Sensing of Environment Vol II: 927-942.

SWAIN, P. H. & S. M. DAVIS (1978): Remote Sensing: The Quantitative Approach. McGraw Hill.

THOMAS, I. L. & G. M. ALLCOCK (1984): Determining the Confidence Level for a Classification. *In:* Photogrammetric Engineering and Remote Sensing, 50(10): 1491-1496.

TOM, C. H. & L. D. MILLER (1984): An Automated Land Use Mapping Comparison of the Bayesian Maximum Likelihood and Linear Discriminant Analysis Algorithms, *In:* Photogrammetric Engineering and Remote Sensing, 50(2): 193-207.

TØMMERVIK, H. (1987): Use of Remote Sensing in Mapping of Vegetation in the Dividalen Area, Central Troms, Northern Norway. *In:* Proceedings of a Workshop on 'Earthnet Pilot Project on Landsat Thematic Mapper Applications', held at Frascati, Italy in December 1987: 271-278.

TOOLPENG, S. (1992): Application of Aerial Photographs in Chalerm Ratanakosin National Park Management Planning, Amphoe Si Sawat Changwat Kanchanaburi. Master Thesis, Kasetsart University, Bangkok, Thailand.

TRSC (1993): Thailand Remote Sensing Center. Distribution Matter of the National Research Council of Thailand. Bangkok. 19.

UENO, S., T. KUSAKA & Y. KAWAKA (1986): Bitemporal Analysis Mapper Data for Land Cover Classification. *In:* Proceedings of IGARSS, 86 Symposium, Zurich, 8-11 September 1986: 523-528.

VANCLAY, J. K. & ROBERT (1990): Utility of Landsat Thematic Mapper Data for Mapping Site Productivity in Tropical Moist Forests. *In:* Photogrammetric Engineering and Remote Sensing, 56(10): 1383-1388.

WACHARAKITTI, S. (1982): Land Use Classification System or Land Use Design. Faculty of Forestry, Kasetsart University, Bangkok. 17. (version Thai)

WALSH, S. J., J. W. COOPER, I. E. VON ESSEN & K. R. GALLAGER (1990): Image Enhancement of Landsat Thematic Mapper Data and GIS Data Integration for Evaluation of Resource Characteristics. *In:* Photogrammetric Engineering and Remote Sensing, 56(8): 1135-1141.

WELCH, R., M. REMILLARD & J. ALBERTS (1992): Integration of GPS, Remote Sensing, and GIS Techniques for Coastal Resources Management. *In:* Photogrammetric Engineering and Remote Sensing, 58(11): 1571-1578.

WOOD, T. F. & G. M. FOODY (1989): Analysis and Representation of Vegetation Continua from Landsat Thematic Mapper Data for Lowland Heaths. *In:* International Journal of Remote Sensing, 10(1): 181-191.

WRIGLEY, R., W. ACEVEDO, D. ALEXANDER, J. BUIS & D. CARD (1984): The Effect of Spatial, Spectral and Radiometric Factors on Classification Accuracy Using Thematic Mapper Data. *In:* Proceedings of IGARSS, 84 Symposium, Strasbourg 27-30 August 1984: 93-100.

WULF, R. R. De. & R. E. GOOSSENS (1987): Classification of Small-Scale Forests in Flanders Using Landsat TM Digital Data : Preliminary Results. *In:* Proceedings of a Workshop on 'Earthnet Pilot Project on Landsat Thematic Mapper Applications', held at Frascati, Italy in December 1987: 241-249.

ZEFF, I. S. & C. J. MERRY (1993) Thematic Mapper Data for Forest Resource Allocation. *In:* Photogrammetric Engineering and Remote Sensing, 59(1): 93-99.

ZHOU, Q. (1989): A Method for Integrating Remote Sensing and Geographic Information Systems. *In:* Photogrammetric Engineering and Remote Sensing, 55(5): 591-596.

APPENDIX A

Landsat TM Geocoded products are supplied in a system-corrected form where systematic errors in both along-line and along-track directions are removed and in precision corrected form using GCP. TM geocoded products are fully compatible with the MSS geocoded products. Each band of this product type comprises of pixels with 25x25 m in size. The product has slightly different dimensions which vary from 2280 to 2360 lines and form 3400 to 3600 pixels per line, depending on the latitude of the product. Each geocoded product is rectangular. Each line of the product contains an equal number of pixels. There are no left fill pixels. Right fill pixels are appended to give a total of 3600 image pixels. In the case of missing image, some of pixels may be black-filled. Adjacent geocoded products contain some common imagery due to the fact that NTS maps are non-rectangular in the UTM projection. There is an overlap of 0-80 pixels and 40-80 lines between adjacent products. The geographic position of a pixel given in UTM or Latitude-Longitude coordinates refers to the top left corner of the pixel.

The data organization by NRCT may be in the form of Band Interleaved by Line (BIL) or Band Sequential (BSQ). We ordered data for use in this study in both forms. In the BIL organization, it contains only one imagery data file, the imagery data for *all* scan lines of one spectral band are grouped together in one imagery data file *before* providing imagery data for the next spectral band in a subsequent file. In both cases, each imagery data file is preceded by a leader file and followed by a trailer data. The leader file for each imagery data file contains scene introductory information, such as the identification, sensor and mission definition, geographic referencing data, processing parameters, and the radiometric transformation tables. Figure 1, 2, 3, and 4 show the file layouts of the Computer Compatible Tapes (CCT).

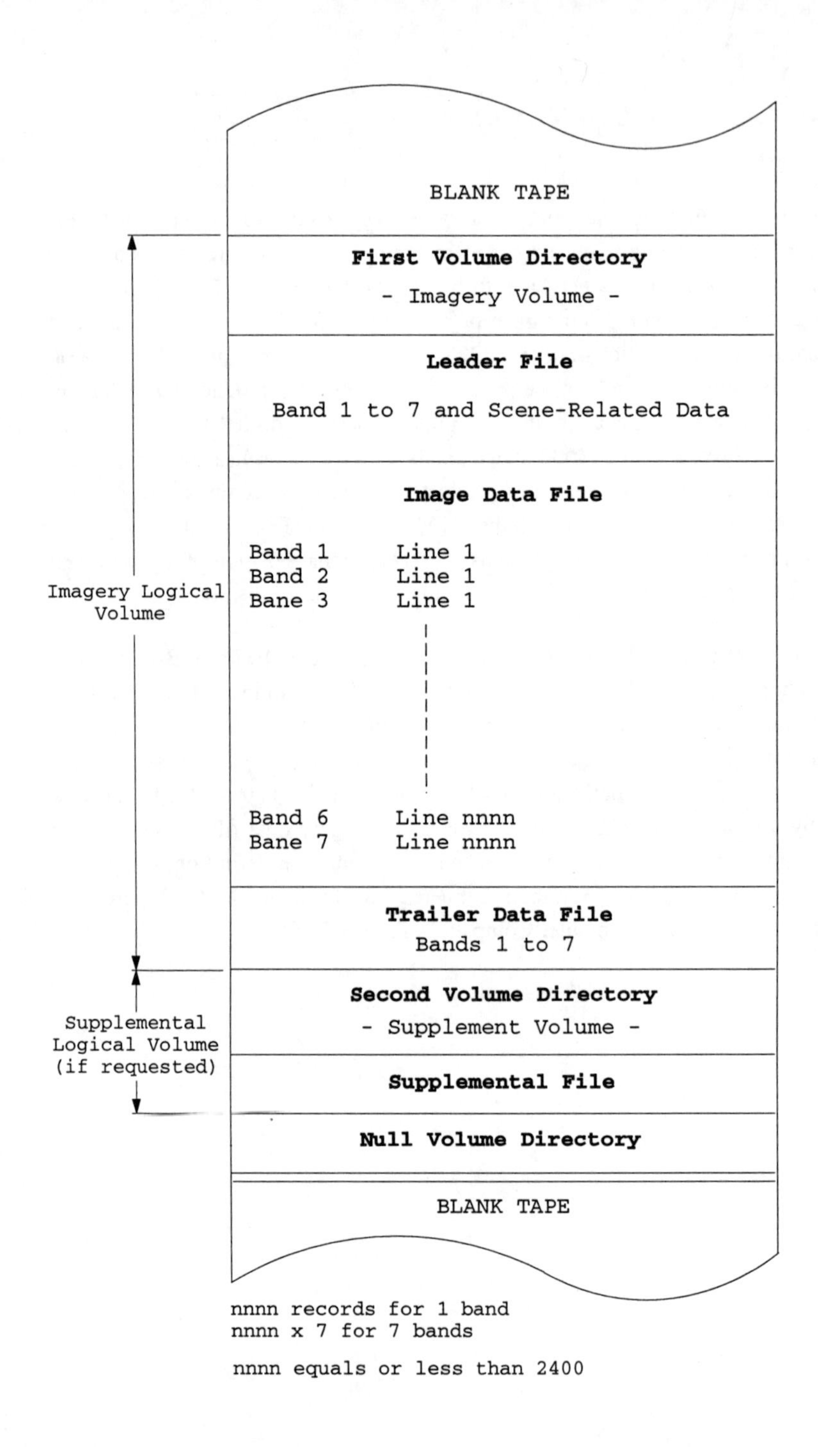

Fig. 1: Geocoded Landsat TM data file layout in CCT with 6250 bpi (BIL).

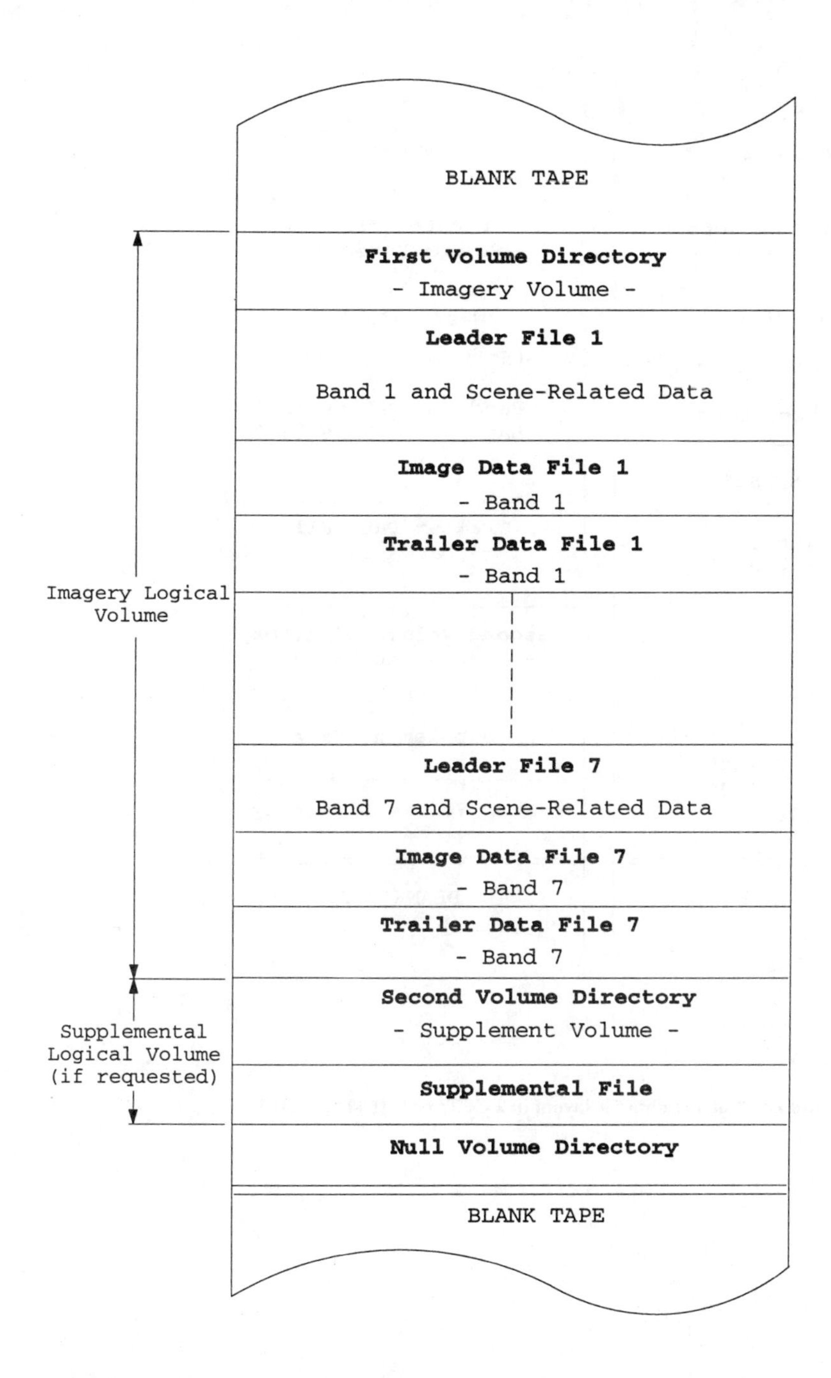

Fig. 2: **Geocoded Landsat TM data file layout in CCT with 6250 bpi (BSQ).**

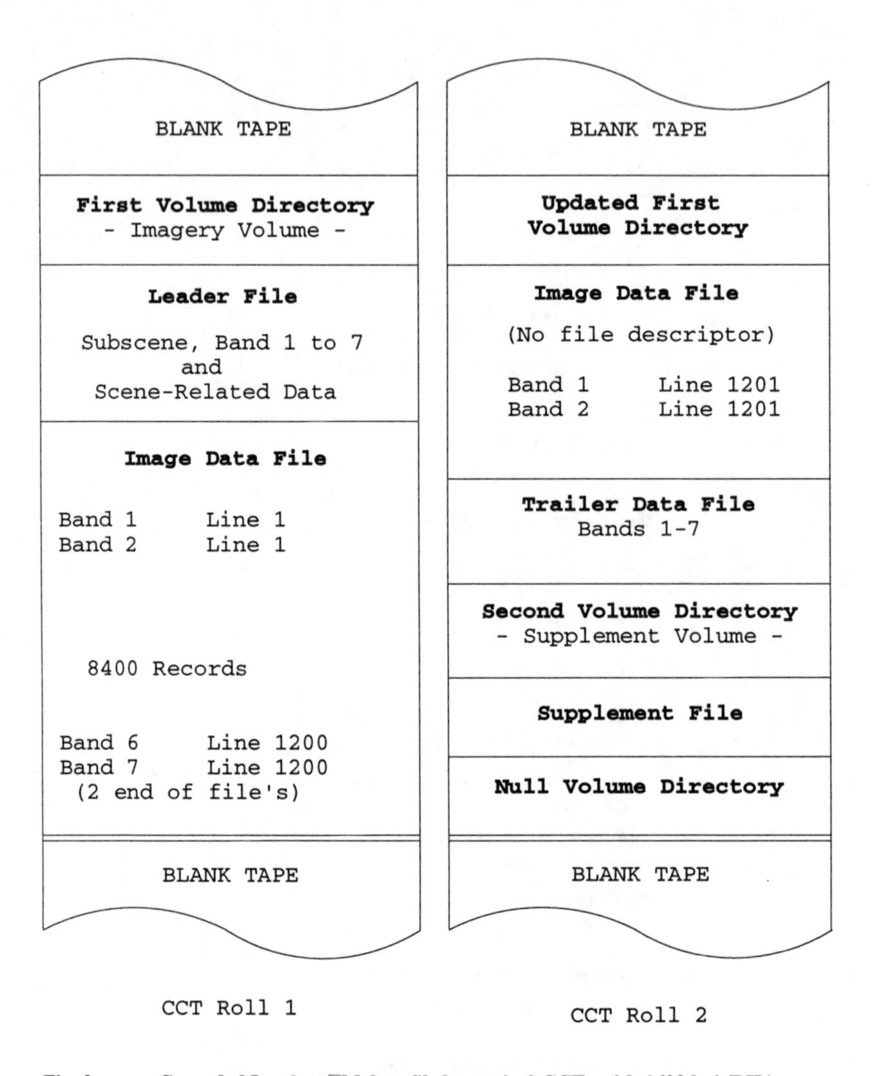

Fig. 3: **Geocoded Landsat TM data file layout in 2 CCTs with 1600 bpi (BIL).**

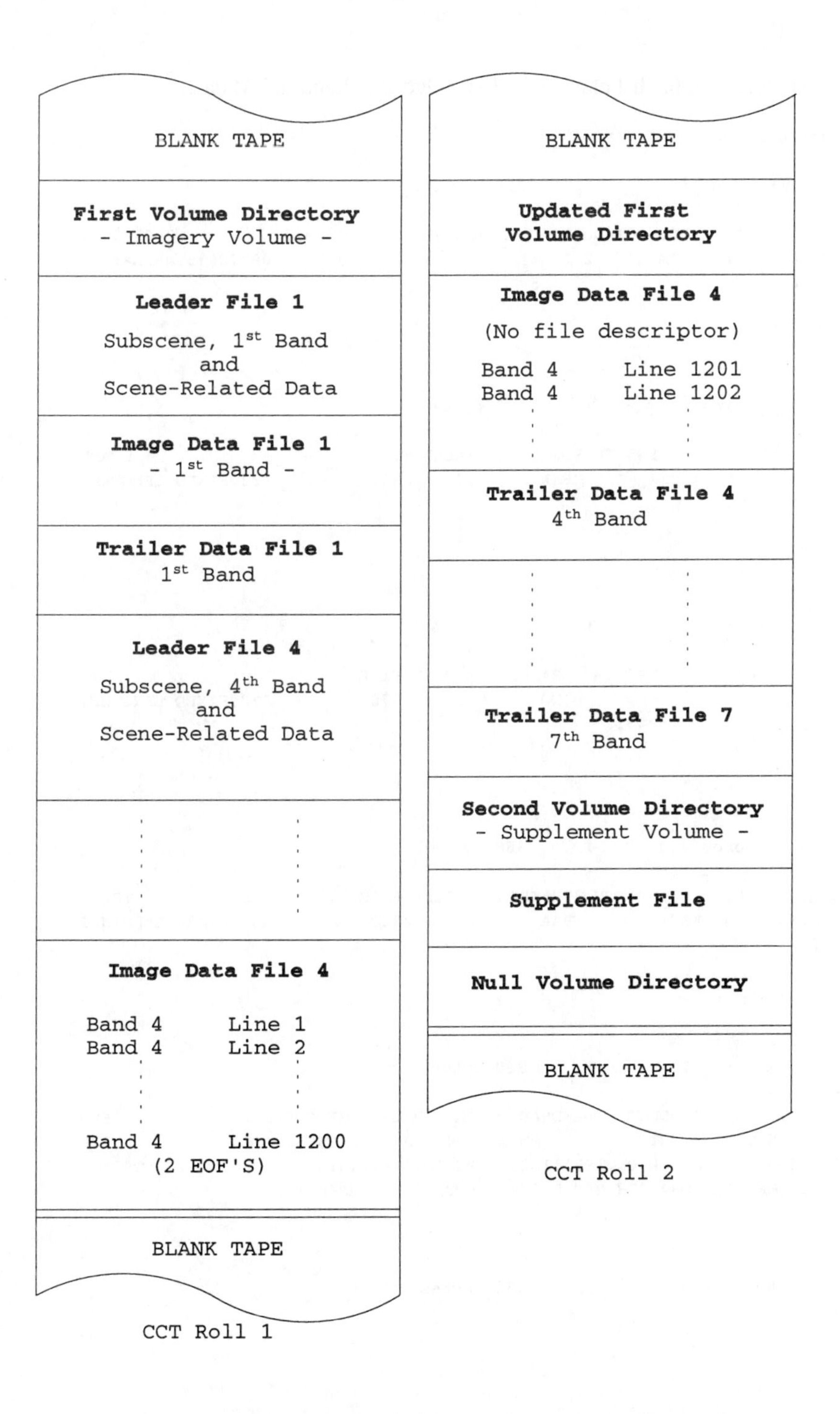

Fig. 4: Geocoded Landsat TM data file layout in 2 CCTs with 1600 bpi (BSQ).

Following is an example of the list of a header file included in Landsat TM data.

```
Tape Structure Listing

File #        1, Records #        1,     360 bytes

00001: ...........hA     CCB-CCT-0002  C  AGICSCVF01     T04810           280203015
00070: 450382 LANDSAT 5  TM      2 1 2 1    1   1     11992042209070635THAILAND
00139:   NRCT-RSOGICS                3   5
00208:
00277:
00346:

File #        1, Records #        2,     360 bytes

00001: ...........hA      1LS5 TM 9LEADBIL LEADER FILE                       LEADM
00070: IXED BINARY AND ASCII      MBAA       17     4320          4320FIXED LENGTHFI
00139: XD 1 1        1       17
00208:
00277:
00346:

File #        1, Records #        3,     360 bytes

00001: ...........hA      2LS5 TM 9LEADBIL IMAGERY FILE                      IMGYB
00070: INARY ONLY                 BINO      8121     3780          3780FIXED LENGTHFI
00139: XD 1 2        1       8401
00208:
00277:
00346:

File #        1, Records #        4,     360 bytes

00001: ...........hA      3LS5 TM 9TRAIBIL TRAILER FILE                      TRAIM
00070: IXED BINARY AND ASCII      MBAA       57     4320          4320FIXED LENGTHFI
00139: XD 2 2        2       0
00208:
00277:
00346:

File #        1, Records #        5,     360 bytes

00001: .....?.....hA   PRODUCT:  LANDSAT  5 TM   BIL7   GEOCODED-PRECIS    9..PRO
00070: CESSED:  NRCT-TRSC GICS        ON 19920422 AT 09070691..SCENE           528
00139: 0203015400      IMAGED ON 19911102..TAPE ID:   T04810                 TAPE
00208:1 OF   2..WRS ID :D942878 MAP 50382 ..LEVEL OF CORRECTION 9..
00277:
00346:

File #        2, Records #        1,     4320 bytes
```

-------------------------------- *and more* --------------------------------

APPENDIX B

Filed notes

Form No. 1
Form No. 2
Form No. 3

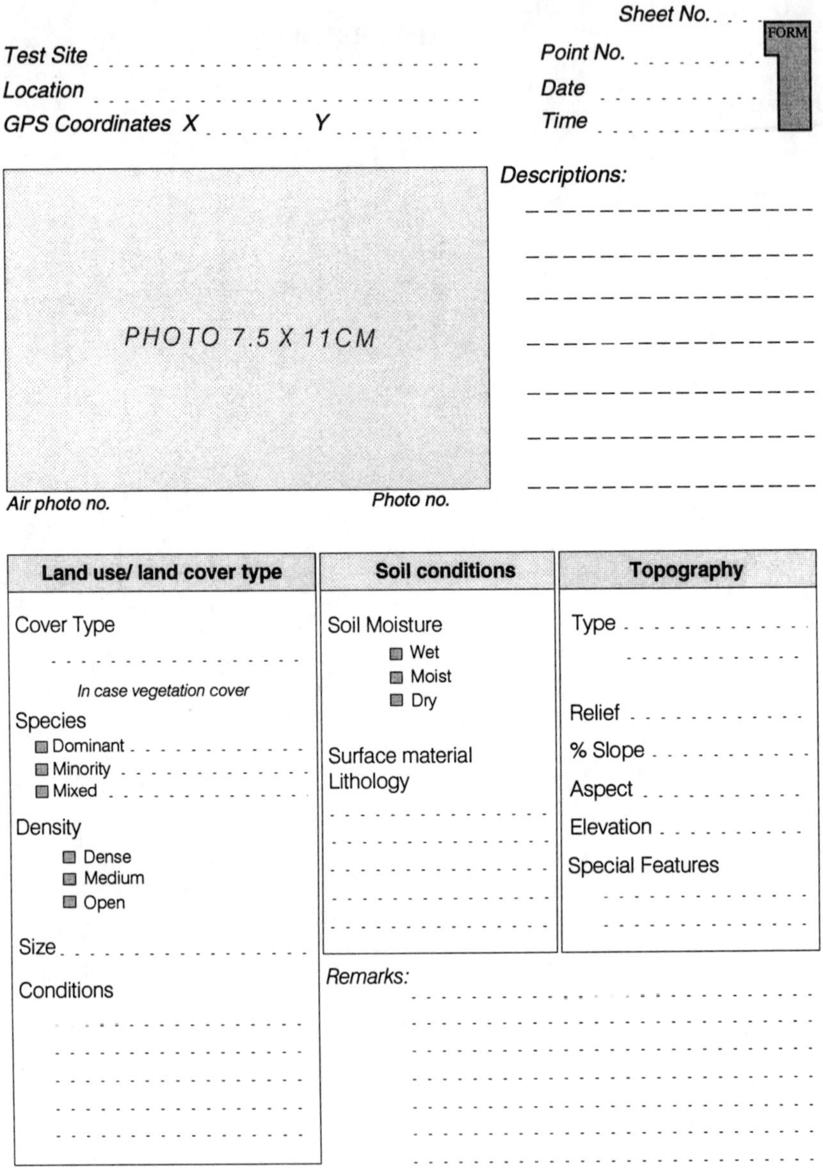

Sheet No. . . . .

FORM 1

Test Site . . . .
Point No. . . . .

Location . . . .
Date . . . .

GPS Coordinates X . . . . *Y* . . . .
Time . . . .

PHOTO 7.5 X 11CM

Air photo no. *Photo no.*

Descriptions:

_ _ _ _ _ _ _ _ _ _ _ _ _ _ _ _ _

Land use/ land cover type	**Soil conditions**	**Topography**
Cover Type . . . . *In case vegetation cover* Species ▪ Dominant . . . . ▪ Minority . . . . ▪ Mixed . . . . Density ▪ Dense ▪ Medium ▪ Open Size . . . . Conditions	Soil Moisture ▪ Wet ▪ Moist ▪ Dry Surface material Lithology	Type . . . . Relief . . . . % Slope . . . . Aspect . . . . Elevation . . . . Special Features . . . .

Remarks:

. . . .

Fig. 1: **Form No. 1 of the field notes used in the field survey.**

CROP CALENDAR

Sheet No. ____

2
FORM

Test Site ______________________________ *Date* ____________

SPECIES	*Month*												*REMARK*
	Jan.	*Feb.*	*Mar.*	*Apr.*	*May*	*Jun.*	*Jul*	*Aug.*	*Sept.*	*Oct.*	*Nov.*	*Dec.*	

Symbol

1 = Planting site preparation stage

2 = Planting stage

3 = Maintenance stage

4 = Harvesting stage

Fig. 2: Form No. 2 of the field notes used in the field survey.

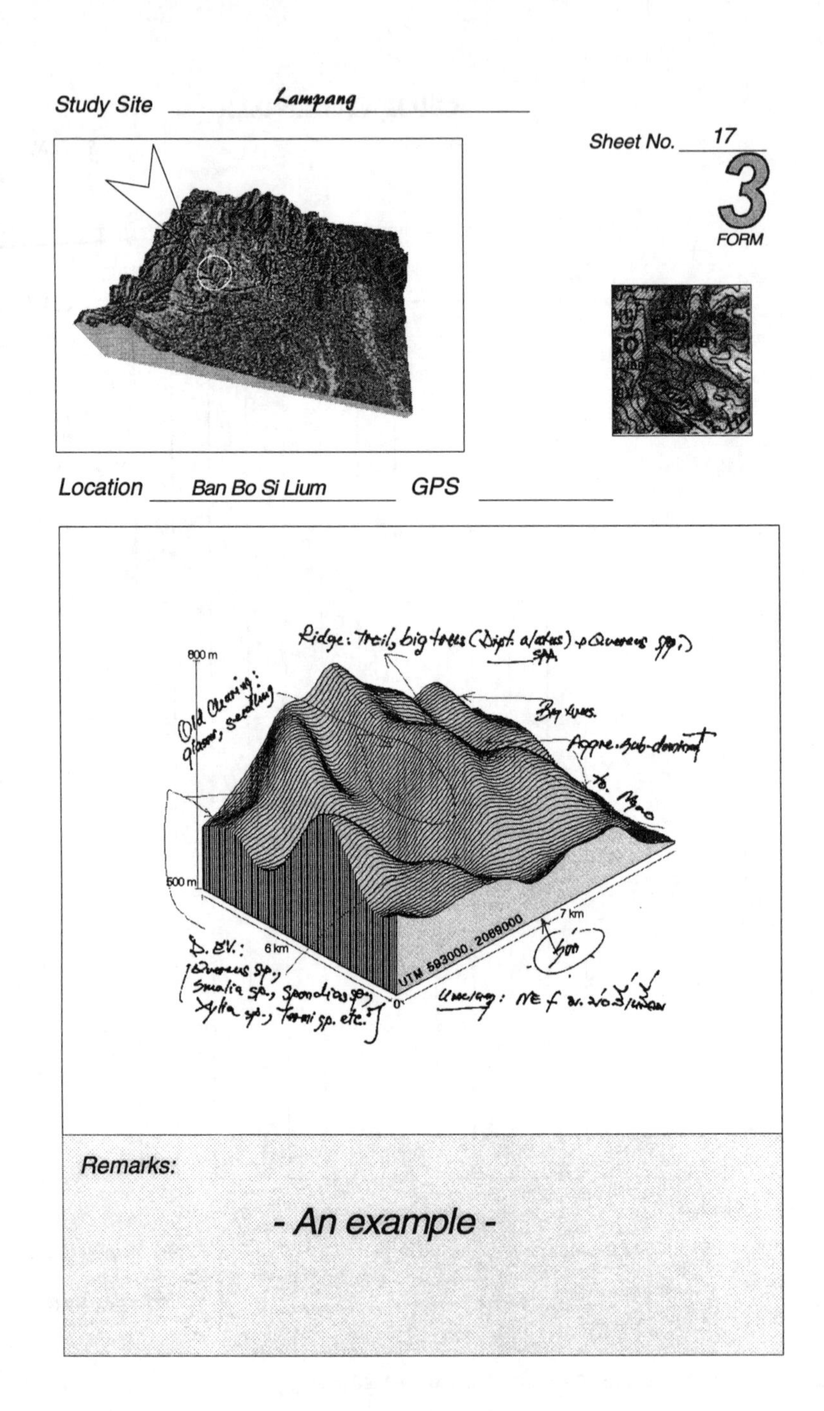

Study Site Lampang

Sheet No. 17

3 FORM

Location Ban Bo Si Lium GPS

Remarks:

- An example -

Fig. 3: Form No. 3 of the field notes used in the field survey.

APPENDIX C

Normally, three-dimensional displays can be done in two planar geometric projection methods, namely parallel projection and perspective projection. Both projection methods calculate the position of each line of an image that is displayed on the view plane (normally, a computer screen). Parallel projection maintains the raster or vectors as a rectangle, in which the image lines remain the same size regardless of the projection in the display as used with 3D correlation histograms. The image is measured to scale but does not give the viewer a sense of proportion. On the other hand, perspective projection foreshortens an image on the screen to give the impression of depth and distance. The lines of an image are projected onto the view plane with the projecting lines converging towards the center of projection. The image cannot be measured to scale but gives the viewer a sense of proportion.

Most of 3D views presented in this study are displayed with the perspective projection method. The following descriptions are briefly summarized in order to help the reader understand the concepts of 3D display.

Figure 1 shows the diagram of perspective elements.

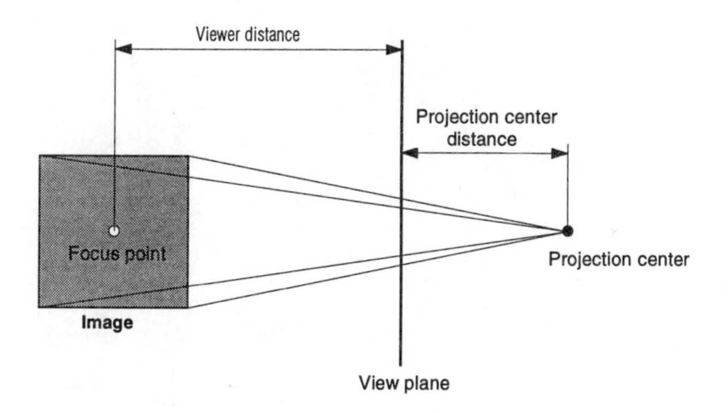

Fig. 1: Diagram of perspective elements.

There are some parameters affect the location of the focus point, the distance between the projection center and the view plane, and the distance between the focus point and the view plane. These parameters are used to control the manner of 3D views. The following parameters are used to control the 3D image displayed in perspective projection:

1. *Rotation angle*. This parameter is used to control the angle of displayed raster or vector objects. The displayed image can be rotated up to 360 degrees around the *Rotation axes*.

2. *Elevation angle*. This parameter is used to affect the movement of the displayed raster or vector object. The elevation angle is an angle formed between the observer's line of sight and ground level. The range of elevation angles is from +90 to -90 degrees.

3. *Z scaling*. This parameter is used to control the vertical exaggeration of an image along the z-axis. The Z scale can be used to adjust an image display from a factor of 0.01 to a number calculated from the minimum and maximum elevation values in the raster object.

4. *Focus point XY*. This parameter is used to control the position of the focus in an X-Y plane for an displayed image. The measurement is in meter.

5. *Focus point Z*. This parameter is used to control the relative Z-dimension of the focus point. The measurement is in meter.

6. *Projection center*. This parameter is used to control the distance from the viewing plane (normally, the screen of a computer) to the projection center. The measurement is in meter.

7. *Viewer distance*. This parameter is used to control the distance from the focus point to the viewing plane for raster display. The measurement is in meter.

The descriptions are based on the TNTmips software functions, released by MicroImages, Inc.

APPENDIX D

In order to take an advantage of the new world wide information technologies using computer networks, ***Internet***, we has installed the so-called a web-browser and a hypertexts server (***WWW Server***) to communicate and keep in touch with the informational highway. **World Wide Web** (WWW) is the protocol of distributed multimedia hyper-documents. These hyper-documents can reside anywhere in Internet's WWW server computers. The documents can contain live links to other documents anywhere with other *Internet Web Servers* over the world. Almost all universities, many government institutes, organizations, and many companies now provide WWW services.

We have been presenting ourselves on the *Internet* through a ***'Home Page'*** (for example Fig. 1 & 2) since early 1995. Our server's address is **http://*Sun-ipx1.bg.TU-Berlin.DE***. Information, in particular images and charts, relevant to this research can be found from this site. It shows the best quality of images and figures which are presented in this book. This would be beneficial for whoever should read this book. Direct URL for an relevant site to this research is:

http://sun-ipx1.bg.tu-berlin.de/pub/staff/kan/Research/

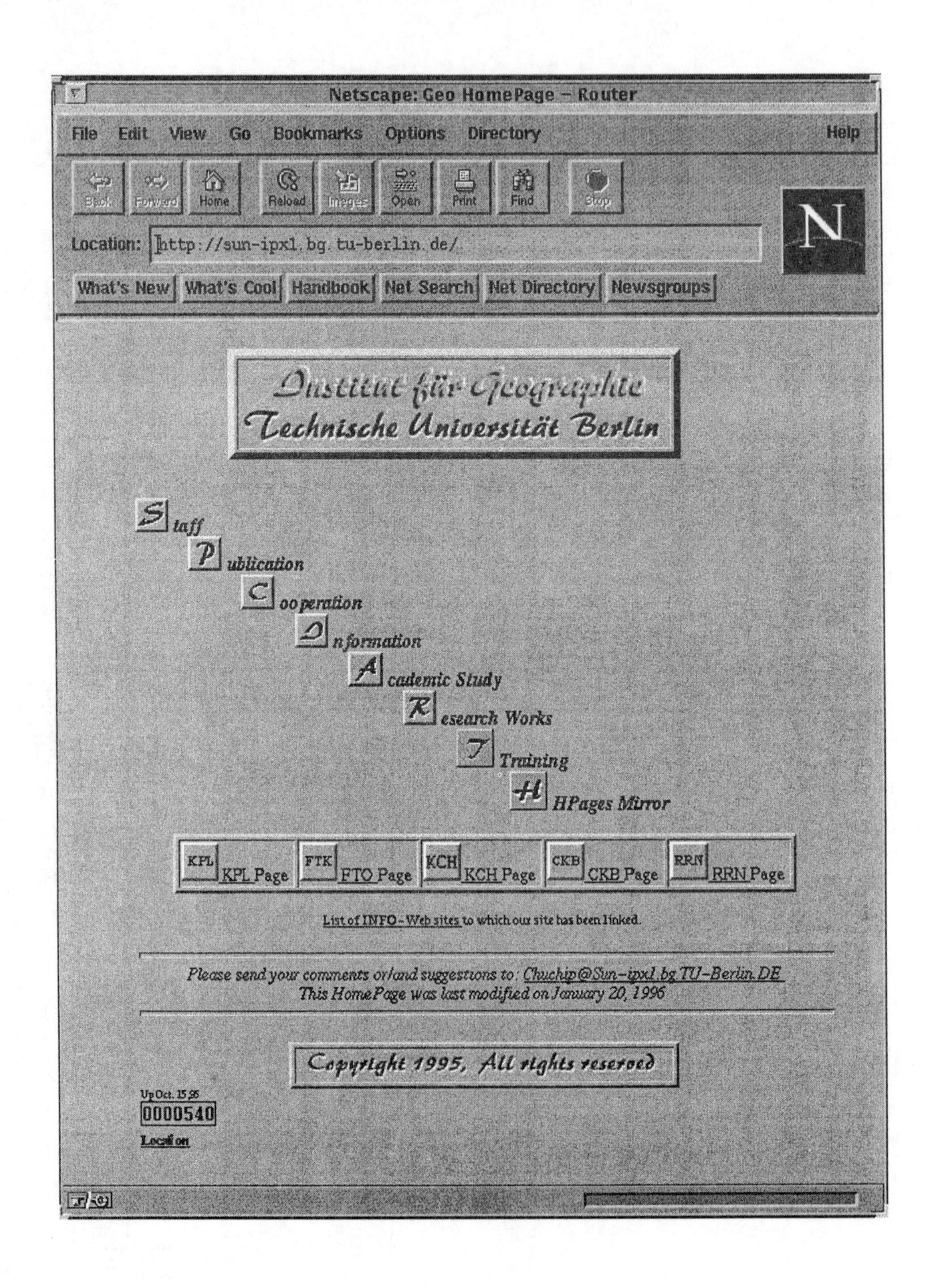

Fig. 1: **A snapshot of Netscape showing the first page of Geo. TU home page.**

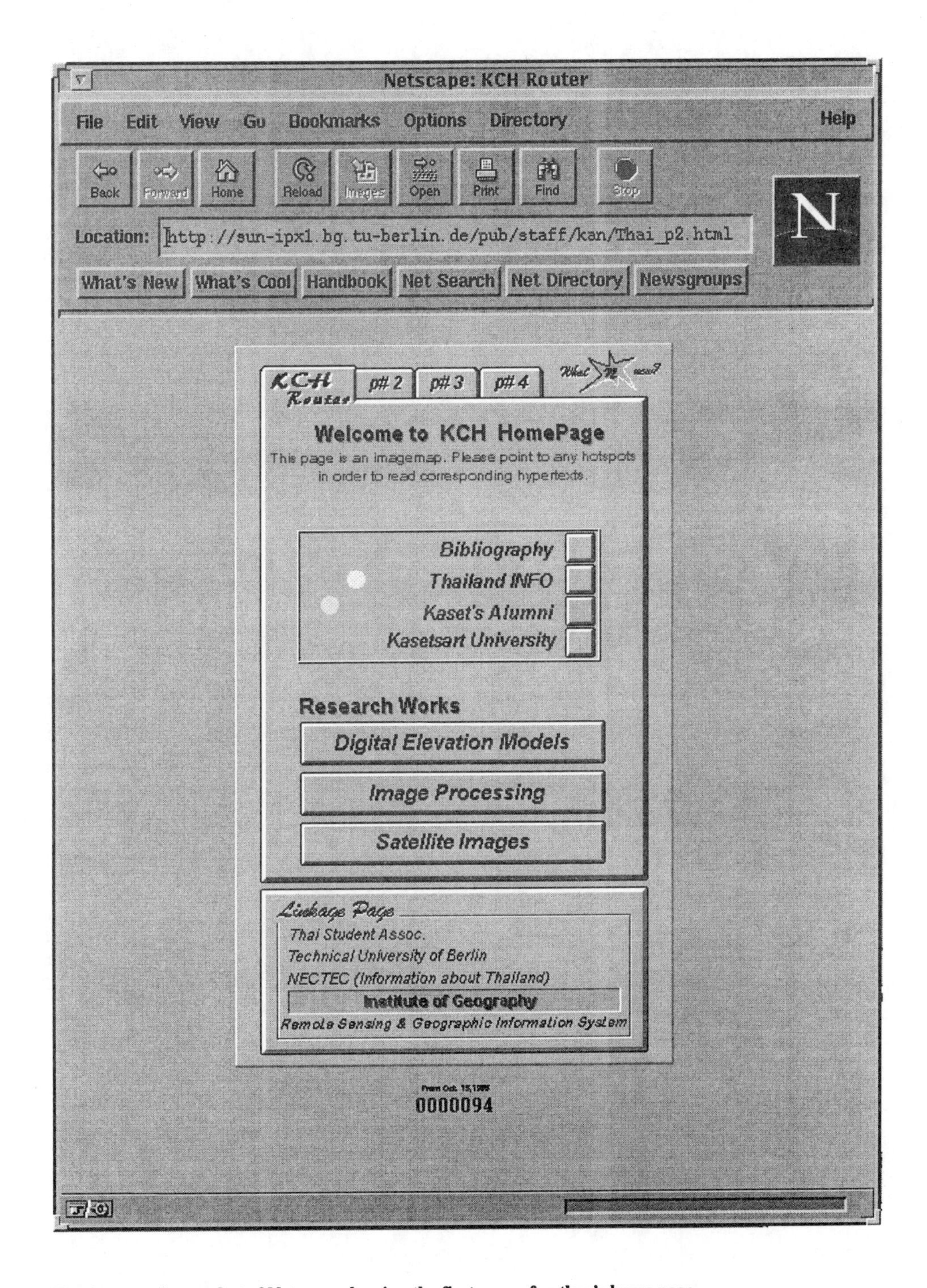

Fig. 2: **A snapshot of Netscape showing the first page of author's home page.**

APPENDIX E

Land use/land cover maps of the study areas
(Attached inner the back cover of the book).

BERLINER GEOGRAPHISCHE STUDIEN

Band 18: HOFMEISTER, Burkhard / VOSS, Frithjof (Hrsg.): Ncue Forschungen zur Geographie Australiens. Ergebnisse aus dem Arbeitskreis Australien. 1986, VI, 180 S., 48 Abb., 15 Tab. und 39 Photos im Text
ISBN 3 7983 1126 9 DM 5,00

Band 19: GABRIEL, Baldur: Die östliche Libysche Wüste im Jungquartär. Vornehmlich nach neueren Feldbefunden. 1986, VI, 216 S., 32 Abb., 114 Tab. und 50 Photos im Text ISBN 3 7983 1132 3 DM 5,00

Band 20: HOFMEISTER, Burkhard / VOSS, Frithjof (Hrsg.): Beiträge zur Geographie der Kulturerdteile. 1986, VII, 346 S., 48 Abb., 46 Tab., 27 Photos und 5 Faltkarten im Anhang
ISBN 3 7983 1133 1 DM 5,00

Band 21: STEINCKE, Albrecht: Freizeit in räumlicher Isolation. Prognosen und Analysen zum Freizeit- und Fremdenverkehr der Bevölkerung von Berlin (West). 1987, XVI, 278 S., 12 Abb., 67 Tab., 19 Photos und 31 Delphi-Übersichten im Text
ISBN 3 7983 1157 9 DM 5,00

Band 22: JASCHKE, Dieter: Die agrarische Tragfähigkeit Australiens. Nutzung und Inwertsetzbarkeit der landwirtschaftlichen Potentiale. 1987, VII, 142 S., 31 Abb. und 42 Tab. im Text
ISBN 3 7983 1158 7 DM 5,00

Band 23: CIMIOTTI, Ulrich (Hrsg.): Beiträge zum Quartär von Holstein. 1987, IV, 212 S., 87 Abb., 2 Photos und 11 Tab. im Text
ISBN 3 7983 1175 7 DM 5,00

Band 24: HOFMEISTER, Burkhard / VOSS, Frithjof (Hrsg.): Neue Forschung zur Geographie Australiens II. Ergebnisse aus dem Arbeitskreis Australien. 1987, VI, 173 S., 44 Abb. und 5 Tab. im Text, 11 S. Photoanhang
ISBN 3 7983 1176 5 DM 5,00

Band 25: HOFMEISTER, Burkhard / VOSS, Frithjof (Hrsg.): Beiträge zur Geographie der Küsten und Meere. Ergebnisse der Symposien Sylt 1986 und Berlin 1987. 1987, VI, 472 S., 167 Abb., 22 Tab. und 53 Photos im Text
ISBN 3 7983 1177 3 DM 5,00

Band 26: SÄNGER, Helmut: Die Vergletscherung der Kap-Karten im Pleistozän. 1988, X, 195 S., 36 Abb., 7 Tab., 15 Karten und 22 Photos im Text bzw. Photoanhang
ISBN 3 7983 1211 7 DM 5,00

Band 27: SCHAAFHAUSEN-BETZ, Sabine: Auswirkungen spontaner Landnahme in Ost-Kalimantan. Untersucht am Beispiel der Srtaße von Samarinda nach Balikpapan Ost-Kalimantan, Indonesien. 1988, VII, 118 S., 24 Abb. (davon 2 Farbkarten), 4 Tab. im Text
ISBN 3 7983 1213 3 DM 5,00

Band 28: SCHULZ, Georg: Lexikon zur Bestimmung der Geländeformen in Karten. 1989, V, 359 S., 296 Abb. incl. 8 farb. Abb. 2. überarbeitete und ergänzte Auflage, 1991
ISBN 3 7983 1283 4 DM 40,00

Band 29: ELLENBERG, Ludwig (Hrsg.): Gefährdung und Sicherung von Straßen in Costa Rica und Panama. 1990, XII, 153 S., 11 Tab. und 63 Karten
ISBN 3 7983 1299 0 DM 10,00

Band 30: GABRIEL, Baldur (Hrsg.): Forschungen in ariden Gebieten. Aus Anlaß der Gründung der Station Bardai (Tibesti) vor 25 Jahren. 1990, VI, 300 S., 10 Tab., 70 Abb., 18 Photos und eine Kartenbeilage
ISBN 3 7983 1340 7 DM 10,00

Band 31: HOFSTEDE, Jacobus: Hydro- und Morphodynamik im Tidebereich der Deutschen Bucht. 1991, X, 113 S., 13 Tab., 41 Abb., 3 Photos im Text
ISBN 3 7983 1422 5 DM 13,00

Band 32: VOLMERG, Rolf-Dieter: Kommunaler Finanzausgleich und zentrale Orte in Schleswig-Holstein. 1991, XVI, 258 S., 87 Tab., 37 Abb. und eine Faltkarte (Tasche)
ISBN 3 7983 1429 2 DM 25,00

Band 33: HOFMEISTER, Burkhard / MÖBIUS, Diana (Hrsg.): Exkursionen durch Berlin und sein Umland. 1992, 404 S.
ISBN 3 9275 7416 3 vergriffen

Band 34: KOLB, Albert: Yünnan - Chinas unbekannter Süden. Mit einem Beitrag von Reinhard Hohler. 1991, XII, 133 S., 7 Tab., 8 Abb. und 30 Photos
ISBN 3 7983 1463 2 DM 22,00

Band 35: SCHÜLLER, Andreas: Zur Morphodynamik des Küstenvorfelds - Innere Deutsche Bucht. 1992, XII, 108 S., 14 Tab., 51 Abb.
ISBN 3 7983 1498 5 vergriffen

Band 36: VOIGT, Bernd: Klima und Landschaft am Horn von Afrika im Quartär. 1992, XII, 151 S., 12 Tab., 46 Abb. und 2 Tafeln
ISBN 3 7983 1499 3 DM 19,00

Band 37: DREISER, Christoph: Mapping and Monitoring of QUELEA Habitats in East Africa. 1993, XII, 149 S., 2 Tab., 90 Abb.
ISBN 3 7983 1560 4 DM 38,00

Band 38: ONGSOMWANG, Suwit: Forest Inventory, Remote Sensing and GIS (Geographic Information System) for Forest Management in Thailand. 1994, XIV, 272 S., 108 Tab. 56 Abb., 20 Photos und 5 Farbkarten (Kartentasche)
ISBN 3 7983 1561 2 DM 63,00

Band 39: NIESTLE, Axel: Drought Risk Modelling in the Nile Valley. Based on a Stream-Aquifer Interaction Model. 1994, X, 81 S., 6 Tab. und 28 Abb.
ISBN 3 7983 1562 2 DM 26,00

Band 40: HOFMEISTER, Burkhard / VOSS, Frithjof (Hrsg.): Exkursionsführer zum 50. Deutschen Geographentag 1995 in Potsdam. 1995, VI, 423 S., 15 Tab, 97 Abb. und 29 Photos
ISBN 3 7983 1641 4 DM 25,00

Band 41: GÖNNERT, Gabriele: Mäandrierung und Morphodynamik in Ästuar am Beispiel der Eider. 1995, XIV, 198 S., 15 Tab., 85 Abb.
ISBN 3 7983 1642 2 DM 55,00

Band 42: ACKER, Heike: Bürobetriebe und Stadtenwicklung. Entwicklungen in Berlin nach 1989 unter besonderer Berücksichtigung der Immobilienbranche. 1995, VII, 172 S., 12 Tab. und 31 Abb.
ISBN 3 7983 1643 0 DM 49,00

Band 43: ALBERS, Christoph: Kommunale Plannung in Alto Valle de Rio Negro y Neaquén, Argentinien. 1996, XII, 244 S., 34 Karten, 21 Abb. und 25 Tab.
ISBN 3 7983 1654 6 DM 49,00

Band 44: STEINECKE, Albrecht (Hrsg.): Stadt und Wirtschaftsraum. Festschrift für Prof. Dr. Burkhard Hofmeister. 1996, XX, 509 S. 41 Tab., 131 Abb. und 16 Photos
ISBN 3 7983 1686 4 DM 28,00

Band 45: ALBERS, Christoph: Planificatión regional en el Alto Valle de Rio Negro y Neaquén, Argentinien. Spanische Version von Band 43. 1996, XII, 245 S., 34 Karten, 21 Abb. und 25 Tab.
ISBN 3 7983 1696 1 DM 49,00

Band 46: CHUCHIP, Kankhajane: Satellite Data Analysis and Surface Modeling for Land Use and Land Cover Classification in Thailand. XIV, 239 S., 96 Abb. (davon 4 Farbkarten), 45 Tab. und 4 Farbkarten (Kartentasche)
ISBN 3 7983 1706 2 DM 49,00

Der Band 34 ist nur bei den Herausgebern zu beziehen.

Nicht ausgeführte Bd-Nrn. sind vergriffen. Bei Abnahme mehrerer Examplare eines Titels wird Preisnachlaß gewährt; Näheres auf Anfrage. Ab 1994 gelten die Preise für den Barverkauf. Bei Bestellungen wird zusätzlich eine Versandpauschale erhoben: für das 1. Exemplar 4,00 DM, jedes weitere Exemplar für 1,00 DM.

Vertrieb/ Publischer: Technische Universität Berlin, Universitätsbibliothek, Abt. Publikationen
Straße des 17. Juni 135, D-10623 Berlin; Tel.: (030) 314-22976, -23676; FAX: (030) 314-24743
Verkauf: Gebäude FRA-B, Franklinstraße 15 (Hof), 10587 Berlin-Tiergarten